Laurier L. Schramm

Innovation Technology: A Dictionary

Also of Interest

Innovation Management.
In Research and Industry
Machado, Davim (Eds.); 2015
ISBN 978-3-11-035872-8, e-ISBN 978-3-11-035875-9

Engineering Risk Management.
2nd Editon
Meyer, Reniers; 2016
ISBN 978-3-11-041803-3, e-ISBN 978-3-11-041804-0

Scientific Leadership.
Niemantsverdriet, Felderhof; 2017
ISBN 978-3-11-046888-5, e-ISBN 978-3-11-046889-2

The Science of Innovation.
A Comprehensive Approach for Innovation Management
Löhr; 2016
ISBN 978-3-11-034379-3, e-ISBN 978-3-11-034380-9

Laurier L. Schramm

Innovation Technology

A Dictionary

DE GRUYTER

Author
Prof. Laurier L. Schramm
The Saskatchewan Research Council
125-15 Innovation Blvd
SASKATOON, Saskatchewan
S7N 2X8
Canada
schramm@src.sk.ca

ISBN 978-3-11-043824-6
e-ISBN (PDF) 978-3-11-042917-6
e-ISBN (EPUB) 978-3-11-042925-1
Set-ISBN 978-3-11-042918-3

Library of Congress Cataloging-in-Publication Data
A CIP catalog record for this book has been applied for at the Library of Congress.

Bibliographic information published by the Deutsche Nationalbibliothek
The Deutsche Nationalbibliothek lists this publication in the Deutsche Nationalbibliografie; detailed bibliographic data are available on the Internet at http://dnb.dnb.de.

Typesetting: Compuscript Ltd., Ireland
Printing and binding: CPI books GmbH, Leck
Cover image: grapestock/iStock/Thinkstock
♾ Printed on acid-free paper
Printed in Germany

www.degruyter.com

Table of Contents

About the Author —— viii
Acknowledgments —— ix

Introduction and Historical Evolution —— 1
Numeric —— 9
Glossary of Terms

A —— 13
B —— 23
C —— 33
D —— 49
E —— 57
F —— 65
G —— 73
H —— 79
I —— 83
J —— 109
K —— 111
L —— 117
M —— 123
N —— 133
O —— 139
P —— 143
Q —— 155
R —— 157
S —— 165
T —— 183
U —— 203
V —— 205
W —— 209
Y —— 213
Z —— 215

References —— 217

Figures

Figure 1. A "*Transilience Map*" Illustrating Abernathy and Clark's Forms of Technological Innovation and Their Influence on an Organization's Prior Technological Knowledge and Resources (horizontal axis) and on the Competitive Marketplace (vertical axis) — **14**
Figure 2. Illustration of Ansoff's Market/Product Matrix. The broken arrow indicates the direction of increasing risk — **18**
Figure 3. Illustration of Henderson and Clark's Forms of Evolutionary Innovation — **20**
Figure 4. Illustration of the "Chain-Linked Model" of innovation, in which feedback loops and alternative developmental pathways are superimposed on a central chain-of-innovation — **35**
Figure 5. Illustration of Technology Readiness Levels and Commercial Readiness Index Levels — **39**
Figure 6. Rogers' Technology Diffusion Model Modified to Illustrate The "*Chasm*," or Tipping Point — **43**
Figure 7. A Simple Technology-Push, Linear Model of Technological Innovation — **51**
Figure 8. Illustration of the "Triple-Helix" Model of a Regional Innovation System — **62**
Figure 9. Illustration of a Stage-Gate® Process for the Development of a New Petroleum Industry Process — **83**
Figure 10. Illustration of a Technology S-curve — **91**
Figure 11. Illustration of an Innovation S-Curve — **91**
Figure 12. Illustration of the Quad-Helix Model of a Regional Innovation System — **93**
Figure 13. Illustration of Innovation Performance Mapping — **98**
Figure 14. Illustration of Innovation Strategy Mapping — **100**
Figure 15. Illustration of Kondratieff Waves and the Industrial Revolutions — **115**
Figure 16. An Interactive Helix-Like Illustration of the SIN Innovation Model — **138**
Figure 17. Illustration of a Family of Technology S-curves — **171**
Figure 18. Illustration of a Technology Hype-Cycle Curve — **189**
Figure 19. Illustration of a "*Universal Success Curve*" — **203**
Figure 20. Illustration of Technology Development Stages and the Valley of Death — **205**

Tables

Table 1. Illustrative Listing of Named Innovation Terms — **3**
Table 2. Some Acronyms Used in the Field of Innovation — **4**
Table 3. Some Famous Names in Innovation — **7**
Table 4. Examples of Innovation Indicators for Regions or Countries — **30**
Table 5. A Generalized Description of Commercial Readiness Index (CRI) Levels — **39**
Table 6. A Simplified Taxonomy for Regional Innovation Systems — **53**
Table 7. Technological Results Produced by Scientific and Engineering Research — **160**
Table 8. Examples of Innovation Indicators for Organizations — **168**
Table 9. A Generalized Description of Technology Readiness Levels (TRLs) — **192**

About the Author

Laurier L. Schramm has over 35 years of R&D experience spanning all four sectors: industry, not-for-profit, government, and academia. He is currently president and CEO of the Saskatchewan Research Council and has previously served as vice president with the Alberta Research Council and president and CEO of the Petroleum Recovery Institute. For much of this time he served in parallel as adjunct professor of chemical and petroleum engineering and before that adjunct professor of chemistry, both with the University of Calgary. His research interests include applied colloid science, interface science and nanoscience. His management interests include applied research, technology development and deployment, and technological innovation.

Dr. Schramm holds 17 patents and has published 13 other books and over 400 other scientific publications or proprietary reports. Many of his inventions have been adopted into commercial practice. He was awarded one of the first NSERC-Conference Board Synergy Awards for Best Practices in University-Industry R&D Partnership, and his work on the development of oil-tolerant foams for enhanced oil recovery was judged to be a Milestone of Canadian Chemistry in the 20th Century by the Canadian Society for Chemistry. He has received other national awards for his work and is a fellow of the Chemical Institute of Canada and an honorary member of the Engineering Institute of Canada.

Among other community service contributions, he served for over two decades on numerous committees and panels of the Natural Sciences and Engineering Research Council of Canada and the Canada Foundation for Innovation, has been a member of several national or international expert advisory panels, and has served on the boards of directors or executive/management committees of numerous other organizations. He is also a co-founder of Innoventures Canada Inc. and a co-founder of Canada's Innovation School.

Acknowledgments

This book evolved out of course notes, and eventually a glossary, that were prepared for offerings of Canada's Innovation School® in Toronto since 2010. Its evolution benefited from the many questions and discussions prompted by the participants in these courses, to whom I am grateful for their enthusiasm, questioning, and interest.

Special thanks to Wanda Nyirfa, Eric Cook, and John McDougall for reading all or parts of early drafts and for their valuable comments and suggestions.

The preparation of this book was made possible through the support of my family, Ann Marie, Katherine, Victoria, and my parents, all of whom have provided consistent encouragement and support.

Even in the modern electronic and Internet age, there remains a need for major research libraries with substantive collections of scientific, engineering, and technical books and periodicals. In the preparation of this book, my work was greatly assisted by the collections of the libraries of the University of Calgary, Carleton University, Massassachusets Institute of Technology (MIT), University of Alberta, University of Saskatchewan, McGill University, and University of Toronto.

Thanks also to the editorial staff of de Gruyter, particularly Karin Sora, Julia Lauterbach, Ria Fritz, Anne Hirschelmann, and Vivien Schubert.

Laurier L. Schramm
December 2016

Introduction and Historical Evolution

The term "*innovation*" comes from the Greek word *kainotomia*, which is derived from *kainos*, or "*new*," and seems to date back to the 5th-century BCE [1]. The term seems to have originally referred to new thoughts, sometimes with a neutral or positive connotation, but more often with a negative connotation [1]. Godin [1, 2] distinguishes between two *episteme*,[1] within which the term "*innovation*" has been understood and used quite differently. In the early modern *episteme*, which held from the 1500s through to the 1800s, innovation meant "*introducing novel change*," particularly with regard to religious and/or political change.

In those early times, the term "*innovation*" seems to have quickly taken on a negative, perjorative connotation. It was frequently meant to imply that the changes were unwanted, unnatural (apart from the natural order of things), revolutionary, and/or dangerous, as in "*introducing change into the established order*" [1]. In this era, whether in the context of politics or religion, introducing changes (innovation), was the purview of the political and religious leaders only. In contrast, the terms "*reformation*" or "*restoration*" were frequently used to describe positive, moderate, natural-order-restoring changes.

The 20th- through 21st-century *episteme*, which held from the 1900s to the present day, brought in the Schumpeterian definition of innovation as introducing novelty to the commercial marketplace through new products and services. This is meant to be a positive, but still disruptive, connotation. The term "technological innovation"[2] is often used to distinguish the Schumpeterian definition of "*innovation*" from that of earlier centuries. Meanwhile, the earlier meaning of the term innovation as introducing novel change has again become widely used in the 21st century, but this time with a positive connotation, and it has been extended to realms beyond those of religion, politics, and technology, such as organizational processes, marketing, and social structures. For example, the Organisation for Economic Co-operation and Development (OECD) recognizes not only product, process, and service innovation, but also marketing innovation and organizational innovation. What in this book is taken to be technological innovation is usually referred to by OECD as "Technological Product and Process Innovation," or "TPP Innovation."

Technological innovation is not new. The conversion of ideas and knowledge into new and commercially successful products, processes, and services has been going on since the beginning of commerce. What are somewhat new are the focus on how to get

---

1 The body of knowledge and understanding that is generally taken to be intellectually certain at any particular time.

2 The term "*technological innovation*" came into use in the 1950s, representing a merger of the work of Maclaurin and Schumpeter and may have been coined by Maclaurin, who frequently referred to "technological change."

more successful product/services into the marketplace when there are so many there already, and the linkage between the introduction of new products, especially "*game-changing*" products into the marketplace and the health, sustainability, and growth potential of entire economies. Systematic studies of innovation practices in the context of economic health, sustainability, or growth do not seem to have occurred until about the time the term *innovation* itself was coined by Schumpeter in the 1930s [3, 4].

Now, more than 70 years since Schumpeter's time, a growing lexicon is associated with innovation because it applies to so many fields of commercial endeavor. The field has also become broadened beyond Schumpeter's original scope because various parties, particularly governments, have focused on ways to help make organizations of all kinds more efficient and/or effective regardless of competitive concerns. This has even extended to the notion of helping to make non-commerical entities more efficient and/or effective, such as not-for-profit organizations (NFPs) including charities, non-government organizations (NGOs), and even government departments and agencies. In this book, the unqualified term "*innovation*" refers to Schumpeter's original definition of innovation, which was exclusively about developing and introducing new commercial products, process, and services into the marketplace [3, 4]. However, in some usage this will be referred-to as "*technological innovation*" in order to make a distinction from "*non-commercial innovation*," which refers to any aspects of improving an organization's efficiency or effectiveness that manifest themselves in ways other than the introduction into the marketplace of new commercial products, process, or services. In this way, the modern, broader uses of the term *innovation*, such as the definition given in the OECD *Oslo Manual*, can be taken to include technological and/or non-commercial innovation. Table 1 illustrates some of the breadth of specifically named "innovation" terms.

This book provides brief explanations for about 1,300 terms and acronyms (Table 2) that may be encountered in a study of the fundamental principles, application approaches and strategies, and commercial aspects of technological innovation. Even this coverage inevitably represents only a personal selection of the terms that could have been included were there no constraints on the size of the book.

I have tried to include as many important terms as possible, and cross-references for the more important synonyms and abbreviations are also included. The difficulty of keeping abreast of the innovation vocabulary has been worsened by the tendency for the language itself to change and broaden since the term "*innovation*" was first coined in the 1930s. Some older terms that are no longer in common use, or worse, that now have completely new meanings, are included as an aid to the reader of the older innovation and economics literature and as a guide to the several meanings that many terms can have. The meaning of the term "*innovation*" itself is still in flux, although some standardization is beginning to occur. I have also included a few terms from the current and possible future waves of economic and technological advances, such as nanotechnology, smart technologies, and robotics, since these tend to recur in writings and conversations about the future of innovation.

Table 1: Illustrative Listing of Named Innovation Terms

20th–21st Century Innovation	Inbound Open Innovation	Innovation System Theory
Adjacent Innovation	Inclusive Innovation	Innovation Trap
Administrative Innovation	Incremental Innovation	Innovation Voucher Program
Adopt and Adapt Innovation	Indigenous Innovation	Innovation Union
Architectural Innovation	Induced Innovation	Innovative Firm
Blowback Innovation	Industrial Innovation	Innovative Industrial Cluster
BOP Innovation	Innovation 2.0	Innovative Regional Cluster
Bottom-Up Innovation	Innovation Activities	Innovator
Catalytic Innovation	Innovation and Sophistication Factors	Institutional Innovation
Chain of Innovation	Innovation at the Bottom of the Pyramid	Integrated Innovation
Closed Innovation	Innovation Barrier	*Jugaad* Innovation
Cognitive Innovation	Innovation Black Box	Knowledge-Induced Innovation
Collaborative Innovation	Innovation Bridge	Lean Innovation
Commcerical Innovation	Innovation Cluster	Local Innovation
Competence-Destroying Innovation	Innovation Continuum	Management Innovation
Competence-Enhancing Innovation	Innovation Deficit	Marketing Innovation
Component Innovation	Innovation Diffusion Model	Market-Pull Innovation
Concept-Push Innovation	Innovation-Driven Economy	Messy Fireworks Innovation
Consumer Innovation	Innovation Ecosystem	Modular Innovation
Continuous Innovation	Innovation Ecosystem Entities	Multiparty Innovation
Core Innovation	Innovation Ecosystem Models	National Innovation System
Cost Innovation	Innovation Expenditures	Needs-Pull Innovation
Customer-Oriented Innovation	Innovation Foresight	Negative Innovation
Demand-Induced Innovation	Innovation Funnel	Non-Commcerical Innovation
Demand-Pull Innovation	Innovation Impacts	Non-Technological Innovation
Design-Driven Innovation	Innovation Incubator	Open Innovation
Discontinuous Innovation	Innovation Inputs	Organizational Innovation
Discovery-Push Innovation	Innovation Intermediary	Other Innovation Activities
Disruptive Innovation	Innovation Killer	Outbound Open Innovation
Downstream Innovation	Innovation Management	Outcome-Driven Innovation
Early Modern Innovation	Innovation Metrics	Participative Innovation
Eco-Innovation	Innovation Models	Platform Innovation
Ecological Innovation	Innovation Outcomes	Political Innovation
Economic Innovation	Innovation Outputs	Pontin's 1st & 2nd Rules of Innovation
Ecosystem Innovation	Innovation Paradoxes	Principles of Innovation
Educational Innovation	Innovation Park	Process Innovation
Evolutionary Innovation	Innovation Pipeline	Product Innovation
Forward Innovation	Innovation Project	Radical Innovation
Fundamental Innovation	Innovation Reach	Recombinant Innovation
Gandhian Innovation	Innovation Sandbox	Regional Innovation Cluster (Hub)
Generations of Innovation	Innovation S-curve	Regional Innovation System
Government Innovation Lab	Innovation Snail	Resource-Constrained Innovation
Hyper-Innovation	Innovation Strategy Mapping	Return on Innovation
Hypo-Innovation	Innovation System	Reverse Innovation
Imitative Innovation	Innovation System Entities	

Table 1 (continued)

Revolutionary Innovation	Socio-Institutional Innovation	TPP Innovation
Routine Innovation	Soft Innovation	Transformational Innovation
Schumpeterian Innovation	Strategic Innovation	Trickle-Up Innovation
Science-Push Innovation	Sustaining Innovation	Upstream Innovation
Self-Determined Innovation	Synthetic Innovation	User Innovation
Service Innovation	Technology-Push Innovation	Waves of Innovation
Shanzhai Innovation	Technological Innovation	*Zizhu Chuangxin*
Social Innovation	Tenets of Innovation	

Table 2: Some Acronyms Used in the Field of Innovation

3M	Minnesota Mining and Manufacturing
ARIZ	Algorithm of Inventive Problems Solving (acronym is for the Russian wording)
ATAR Model	Awareness, Trial, Availability, Repeat Model
BERD	Business Enterprise Expenditure on R&D
BI	Business Intelligence
BOP Innovation	Innovation at the Bottom of the Pyramid
BRIC(S/K)	Brazil, Russia, India, China, and South Korea
CapEx	Capital Expense
CDA	Confidential Disclosure Agreement. *See* Non-Disclosure Agreement
CE	Concurrent Engineering.
CI	Competitive Intelligence
CRI	Commercial Readiness Index
CRISPR	Clustered Regularly Interspaced Short Palindromic Repeats
CTEs	Critical Technology Events
CVCA	Customer Value-Chain Analysis
DFA	Design for Assembly. *See* Design for Excellence.
DFM	Design for Manufacture. *See* Design for Excellence.
DFMA	Design for Manufacture and Assembly. *See* Design for Excellence.
DFX	Design for Excellence.
DHL GCI	DHL Global Connectedness Index. *See* Connectedness Index.
DIKW Hierarchy	Data-Information-Knowledge-Wisdom Hierarchy. *See* Wisdom Hierarchy
DPI	Domestic Product of Industry
EBIT	Earnings before interest and taxes. *See* Earnings Before Interest, Taxes, Depreciation, and Amortization
EBITA	Earnings before interest, taxes, and amortization. *See* Earnings Before Interest, Taxes, Depreciation, and Amortization
EBITD	Earnings before interest, taxes, and depreciation. *See* Earnings Before Interest, Taxes, Depreciation, and Amortization
EBITDA	Earnings Before Interest, Taxes, Depreciation, and Amortization
EBITDAR	Earnings before interest, taxes, depreciation, amortization, and restructuring or rent costs. *See* Earnings Before Interest, Taxes, Depreciation, and Amortization
FBE	Fuzzy Back-End

Table 2 (continued)

FEF	Front-End Fuzziness. *See* Fuzzy Front-End
FFE	Fuzzy Front-End
FFF Capital	Friends, Family, and Fools Capital. *See* Seed Capital
FTE	Full-Time Equivalent
GAAP	Generally Accepted Accounting Principles
GBAORD	Government Budget Appropriations or Outlays for R&D. *See* Government R&D Expenditures
GCI	Global Competitiveness Index
GDP	Gross Domestic Product
GERD	Gross Domestic Expenditure on R&D
GIL	Government Innovation Lab
GNERD	Gross National Expenditure on R&D
GNP	Gross National Product
GOVERD	Government Intramural Expenditure on R&D
HERD	Higher Education Expenditure on R&D
I18N	"I-eighteen letters-N." *See* Internationalization (Product, Process, or Service).
ICT	Information and Communication Technology
ILO	Industry Liaison Office
IoT	Internet of Things
IP	Intellectual Property
I-Pipe	Innovation Pipeline
IP Mining	Intellectual Property Portfolio Mining
IP Portfolio Mining	Intellectual Property Portfolio Mining
IPR	Intellectual Property Rights
ISF	Innovation System Foresight. *See* Foresight
KBE	Knowledge-Based Economy
KETs	Key Enabling Technologies
KIBS	Knowledge Intensive Business Services
KIS	Knowledge Intensive Services. *See* Knowledge Intensive Business Services
KM	Knowledge Management
K-Waves	Kondratieff Waves
L10N	"L-ten letters-N." *See* Internationalization (Product, Process, or Service)
LEPEST Analysis	Legal, Environmental, Political, Economic, Social and Technological Analysis. *See* Social, Technological, Economic, Environmental and Political Analysis
ME	Medium-Sized Enterprise. *See* Small- and/or Medium-Sized Enterprise
MER	Mandate Effectiveness Ratio
MFP	Multifactor Productivity
MGI CI	McKinsey Global Institute Connectedness Index. *See* Connectedness Index
MNC	Multinational Enterprise
MNE	Multinational Enterprise
MSTI	Main Science and Technology Indicators
MTA	Material Testing (or Transfer) Agreement. *See* Material Testing Agreement
NDA	Non-Disclosure Agreement
NGO	Non-Government Organization

Table 2 (continued)

NIH Syndrome	Not Invented Here Syndrome
NIS	National Innovation System. *See* Innovation Ecosystem
NPD	New Product Development
NPVI	New Product Vitality Index
NSI	National System of Innovation. *See* Innovation Ecosystem
OECD	Organisation for Economic Co-operation and Development
OEM	Original Equipment Manufacturer
OpEx	Operating Expense. *See* Capital Expense
PACE Process	Product and Cycle-Time Excellence Process
P&ID	Piping and Instrumentation Diagram
PEST Analysis	Political, Economic, Social, and Technological Analysis. *See* Social, Technological, Economic, Environmental, and Political Analysis
PESTEL Analysis	Political, Economic, Social, Technological, Environmental, and Legal Analysis. *See* Social, Technological, Economic, Environmental, and Political Analysis
PESTLE Analysis	Political, Economic, Social, Technological, Legal, and Environmental Analysis. *See* Social, Technological, Economic, Environmental, and Political Analysis
PFD	Process Flow Diagram
P-KIBS	Professional Knowledge Intensive Business Services
PLC	Product Life-Cycle. *See* S-curve
pMTA	Plant Material Testing Agreement. *See* Material Testing Agreement
R&D	Research and Development
RCI	Resource-Constrained Innovation.
RIS	Regional Innovation Systems
ROI	Return on Investment
ROI2	Return on Innovation Investment. *See* Return on Innovation
ROII	Return on Innovation Investment. *See* Return on Innovation
ROInn	Return on Innovation
RTO	Research and Technology Organization
SG&A Expenditures	Selling, General, and Administrative Expenditures. *See* Capital Expense
SIN Innovation Model	Systems Integration and Networking Model of Innovation. *See* Non-Linear Innovation Models
SLEPT Analysis	Social, Legal, Economic, Political, and Technological Analysis. *See* Social, Technological, Economic, Environmental, and Political Analysis
SMB	Small- and Medium-Size Business. *See* Small- and Medium-Size Enterprise
SmE	Small-Sized Enterprise. *See* Small- and/or Medium-Sized Enterprise
SME	Small- and/or Medium-Sized Enterprise
SROI on R&D	Social Return on Investment in R&D
STEEP Analysis	Social, Technological, Economic, Environmental, and Political Analysis
STEEPLE Analysis	Social, Technological, Economic, Environmental, Political, Legal, and Ethical Analysis. *See* Social, Technological, Economic, Environmental, and Political Analysis
STEEPLED Analysis	Social, Technological, Economic, Environmental, Political, Legal, Ethical, and Demographic Analysis. *See* Social, Technological, Economic, Environmental, and Political Analysis

Table 2 (continued)

STEEPV Analysis	Social, Technological, Economic, Ecological, Political, and Values Analysis.
STEER Analysis	Social, Technological, Economic, Ecological, and Regulatory Analysis *See* Social, Technological, Economic, Environmental, and Political Analysis
STEP Analysis	Social, Technological, Economic, and Political Analysis. *See* Social, Technological, Economic, Environmental, and Political Analysis
SWOT Analysis	Strengths, Weaknesses, Opportunities, and Threats Analysis
TAM	Technology Acceptance Model
TH Theory	Triple-Helix Theory. *See* Triple-Helix Model
TIPS	Theory of Inventive Problem Solving
T-KIBS	Technological KIBS. *See* Knowledge Intensive Business Services.
TMI	Technology Market Intermediaries
TPP	Technological Product and Process. *See* Innovation (20th–21st century)
TRI	Technology Readiness Index
TRIZ	Theory of Inventive Problem Solving (acronym is for the Russian wording)
TRL	Technology Readiness Level
TTM	Time to Market
UILO	University-Industry Liaison Office. *See* Industry Liaison Office
USPTO	U.S. Patent and Trademark Office
VoC	Voice of the Customer
WEF	World Economic Forum. *See* Global Competitiveness Index, Competitiveness Drivers

Table 3: Some Famous Names in Innovation

3M Inc.	10
Altshuller, Genrich	17
Bush, Vannevar	26
Chesbrough, Henry W.	35
Christensen, Clayton	35
Cooper, Robert	43
Drucker, Peter	55
Etzkowitz, Henry	62
Foster, Richard	69
Holland, Maurice	81
Kondratieff, Nikolai	114
Maclaurin, (William) Rupert	123
Porter, Michael	148
Rogers, Everett	163
Schmookler, Jacob	166
Schumpeter, Joseph	167
Solow, Robert	177
Verhulst, Pierre	207

I have also included a selection of brief biographical introductions to some of the pioneers in the field (see Table 3).

Specific literature citations are given when the sources for further information are particularly useful, unique, or difficult to find. Some particularly helpful sources include references [5, 6, 7, 8]. There are a few other published dictionaries or glossaries of innovation, such as references [2, 9, 10]. Any or all of the above can be good starting points for further information.

Numeric

10/5 Rule	A venture capital "rule of thumb" that a commercialized product, process, or service should yield a 10-times return on investment within a period of 5 years. *See also* Venture Capital.
1st-Generation Innovation Model	The Technology-Push Model. *See* Generations of Innovation, Linear Innovation Models.
1st-Generation Nanotechnology	*See* Generations of Nanotechnology.
1st Generation of Technology Foresight	*See* Generations of Technology Foresight.
1st Industrial Revolution	*See* Technological Ages.
1st Wave	The first Kondratieff Wave, spanning the industrial revolution era and lasting from about 1780 to about 1830. Key developments during this wave included the steam engine and general industrialization. Reference [11]. *See also* Kondratieff Waves.
20th–21st-Century Innovation	*See* Innovation (20th–21st century).
20/30 Rule	A market potential "rule of thumb" that an invasionary technology should perform its function 20% better and 30% cheaper than the pre-existing, competing technologies. *See also* Best Available Technology. Reference [6].
2nd-Generation Innovation Model	The Market-Pull Model, *See* Generations of Innovation, Linear Innovation Models.
2nd-Generation Nanotechnology	*See* Generations of Nanotechnology.
2nd Generation of Technology Foresight	*See* Generations of Technology Foresight.

DOI 10.1515/9783110429176-001

2nd Industrial Revolution	*See* Technological Ages.
2nd Wave	The second Kondratieff Wave, spanning the industrial production era and lasting from about 1830 to about 1880. Key developments during this wave included railways, steel, and heavy engineering. Reference [11]. *See also* Kondratieff Waves.
3M Inc. (1902 – Present)	An American company founded in 1902 as the Minnesota Mining and Manufacturing Company. The company's original foci were on mining for grinding-wheel abrasives and later sandpaper products. Over time, 3M developed innovation as a core competency, leading to a number of famous products, including Post-it® notes, Scotch-Brite™ products, and Magic™ tape. 3M has continued to be a champion and a leader in making innovation central to its business strategy. At one point, 3M had a goal of having 30% of total sales be from new products developed and commercialized within the previous 5 years. Reference [12]. *See* New Product Vitality Index.
3rd-Generation Innovation Model	The Coupling Model, *See* Generations of Innovation, Non-Linear Innovation Models.
3rd-Generation Nanotechnology	*See* Generations of Nanotechnology.
3rd Generation of Technology Foresight	*See* Generations of Technology Foresight.
3rd Industrial Revolution	*See* Technological Ages.
3rd Wave	The third Kondratieff Wave, spanning the scientific revolution era and lasting from about 1880 to about 1930. Key developments during this wave included electricity, chemistry, and the chemical industry. Reference [11]. *See also* Kondratieff Waves.
4th-Generation Innovation Model	The Integrated Model. *See* Generations of Innovation, Non-Linear Innovation Models.
4th-Generation Nanotechnology	*See* Generations of Nanotechnology.

4th Generation of Technology Foresight	*See* Generations of Technology Foresight.
4th Industrial Revolution	*See* Technological Ages.
4th Pillar Organization	*See* Fourth-Pillar Organization.
4th Wave	The fourth Kondratieff Wave, spanning the technical revolution era and lasting from about 1930 to about 1970. Key developments during this wave included automobiles, mass production, and the petrochemical industry. Reference [11]. *See also* Kondratieff Waves.
5th-Generation Innovation Model	The Systems Integration and Networking Model. *See* Generations of Innovation, Non-Linear Innovation Models.
5th Generation of Technology Foresight	*See* Generations of Technology Foresight.
5th Wave	The fifth Kondratieff Wave, spanning the information and telecom revolution era and lasting from about 1970 to about 2010. Key developments during this wave included microcomputers, information technology, and telecommunications technology. Reference [11]. *See also* Kondratieff Waves.
6th Wave	The sixth Kondratieff Wave, spanning the current era and predicted to last from about 2010 to about 2050. Key developments during this wave might include environmental technology, genetic engineering, nanotechnology, robotics, and health technologies. Reference [11]. *See also* Kondratieff Waves.
7th Wave	The predicted seventh Kondratieff Wave, spanning a future era and predicted to last from about 2050 to about 2090. It has been speculated that the seventh wave may be driven by a merging of technology and intelligence to create autonomous robots, whether at nanoscale, microscale, or macroscales, that are capable of independent action, self-repair, and replication. Reference [11]. *See also* Kondratieff Waves.

A

Abernathy-Clark Model An innovation model that defines forms of evolutionary innovation and distinguishes between an innovation's effect on organization's technological knowledge and resources and its effect on the scale of the technological advance and whether the competing products, processes, or services remain somewhat competitive or are made obsolete. In the Abernathy-Clark Model, organizations that simply enhance their technological knowledge and resources are most likely to achieve modest incremental innovation at best (they termed it "regular innovation"), whereas those that develop completely new technological knowledge and/or resources and use them to make huge, game-changing (i.e., market-changing) technological advances, are most likely to achieve disruptive innovation (they termed it "architectural innovation"). In between these extremes fall two categories of organizations. One category represents organizations that develop completely new technological knowledge and/or resources but only use them to make more competitive products that do not displace or render obsolete their competing products. Such companies achieve evolutionary innovations (they termed it "revolutionary innovation"). The other category represents organizations that develop advances in their technological knowledge and/or resources and are able to introduce products that displace or render obsolete their competing products in a small market niche. Such companies achieve niche innovation. The Abernathy-Clark Model has been used to explain how incumbent companies tend to be in good position to implement incremental innovations, as they can leverage their existing technological knowledge and resources, whereas new companies entering a marketplace tend to be in a good position to implement disruptive innovations, as they do not need to change their existing technological knowledge and/or resources in order to do so. Reference [13]. *See* Figure 1. *See also* Incremental Innovation, Evolutionary Innovation, Disruptive Innovation, Henderson-Clark Model.

DOI 10.1515/9783110429176-002

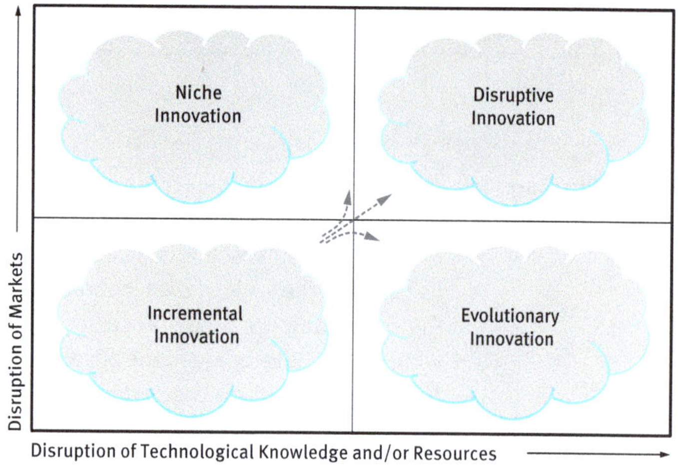

Figure 1: A "*Transilience Map*" Illustrating Abernathy and Clark's Forms of Technological Innovation and Their Influence on an Organization's Prior Technological Knowledge and Resources (horizontal axis) and on the Competitive Marketplace (vertical axis).

ABG Model — *See* Triple Helix Model.

Ab Initio — In science and engineering, the Latin term *ab initio* generally refers to developing models and/or making calculations based on first principles; without experimental data (other than fundamental physical constants).

Absorptive Capacity — *See* Technology Acceptor Capacity.

Academic Revolution — *See* Second Mission, Entrepreneurial University.

Accelerator — *See* Business Accelerator.

Acceptance Threshold — The transition realm of customer perceptions, between considering a technology to be unacceptable (or unattractive, or not useful) and acceptable (or attractive, or useful). Unacceptable technologies can sometimes be turned into acceptable technologies through an "*itemized response*" process of analyzing each negative or unacceptable feature (whether real or perceived) of a technology and finding ways to change or address each such drawback. Also termed Threshold of Acceptance. *See also* Technology Readiness Index, Technology Acceptance Model.

Acquisition of Technology — The purchase of, or barter for, externally derived knowledge of how to effectively use a product, process, or service (know-how), a way of conducting or controlling a manufacturing activity (a practice or process), or a thing to be manufactured, used, or consumed (product). Two broad categories are sometimes distinguished in this context: embodied and disembodied technology. Embodied technology refers to technology that is acquired indirectly because it is embedded in machines, instruments, new employees, or contractors that incorporate or contain the knowledge. Disembodied technology refers to technology that resides in trade secrets, books and other publications, patents, licenses, trademarks, software, and the like. *See also* Knowledge, Embedded Knowledge.

Active Materials — *See* Smart Materials.

Activities — (Innovation) Any or all scientific, engineering, technological, organisational, financial, or commercial activities that lead, or are intended to lead, to development and deployment of innovation(s). Innovation activities include R&D plus subsequent activities such as experimental development, prototyping, pilot testing, demonstration, and preproduction. *See also* Inputs, Outcomes, Outputs, Reach, Research and Development, Metrics, Impacts. Reference [6].

Adapt and Adopt — *See* Adopt and Adapt.

Adjacent Innovation — The commercialization of a product, process, or service that is already being sold in one market into a new market, particularly where the new market is in some sense “near” the established market. This term is sometimes alternatively used as a synonym for *incremental innovation*, in the sense of a minor improvement to a product, process, or service in an existing market. *See also* Incremental Innovation, Disruptive Innovation, Evolutionary Innovation.

Administrative Innovation — A kind of *Organizational Innovation*, administrative innovation refers to new knowledge applied to developing and implementing new and improved organizational structures and administrative processes, including business strategy. Also termed *Business Process Innovation* or *Business Structure Innovation*, respectively. Such new organizational processes should have some kind of efficiency or productivity benefit, even if they are internal and have little or no connection to technological innovation. Example: The Total Quality Management processes pioneered by W. Edwards Deming in the United States in the 1980s. A third form is *Business Culture Innovation*, in which an organization sets about to change its internal culture in order to improve its organizational performance. *See also* Organizational Innovation, Innovation.

Adopt and Adapt	A technology development and/or innovation strategy in which one adopts an existing technology from another business, industry or market and adapts it to the innovator's needs. Also termed Adapt and Adopt. *See also* Ansoff Matrix, Fast Follower.
Adoption	*See* Technology Adoption Lifecycle.
Advanced Technology	*See* High-Technology.
Advanced Technology Development	*See* Research and Development.
Agency Theory	*See* Theory of the Firm.
Agile Development	A product development approach that involves collaboration among self-organizing, cross-functional teams. This approach facilitates adaptive planning, evolutionary development, responsiveness, and continuous improvement. The "*agile development*" concept seems to have originated in the computer software industry.
Agnostic Marketing	A term coined by Clayton Christensen, referring to the inability to know in advance whether, how, or in what quantities customers will use a disruptive innovation product until some first customers try using it. This suggests maintaining a very broad (i.e., agnostic) view of what the ultimately successful market could be. Reference [14].
Algorithm of Inventive Problems Solving	(ARIZ is the Russian acronym) One of the tools used to try to solve invention problems in Altshuller's Theory of Inventive Problem Solving. *See* Altshuller, Creative Thinking Models, Theory of Inventive Problem Solving.

Alliance Capitalism — The trend for some groups of companies to form cooperative relationships and/or strategic alliances in order to accelerate their technological innovation processes and to penetrate international markets. The alliance companies can include partners, suppliers, distributors, and even competitors.

Alpha Release — *See Alpha* Test.

Alpha Test — One of several levels of pre-release testing conducted as part of the development of a new product, process, or service. References [15, 16].

"*Alpha Testing*" is usually conducted internally, once all of the intended features have been built in, and usually includes testing for all aspects of integration and performance. If this version passes the testing, it is referred to as an "*Alpha Release*" of the product.

"*Beta Testing*" is conducted by a selected number of external end-users. This is considered to be pre-release testing and provides a mechanism for obtaining end-user feedback while providing the marketplace with a preview of the intended product. This version is referred to as a "*Beta Release*" of the product.

"*Gamma Testing*" is sometimes conducted by customers in a limited market with a product that is almost but not quite ready for full market release. This is somewhat like a next level of *Beta Testing* and may be aimed at testing a specific feature of the product, such as an aspect of product safety, before the full release. This version is referred to as a "*Gamma Release*" of the product.

Altov, Genrikh — *See* Altshuller.

Altov, Henry — *See* Altshuller.

Altshuller, Genrich (Saulovich) (1926 – 1998) — A Soviet mechanical engineer and inventor best known in the innovation world for his *Theory of Inventive Problem Solving*, a systematic approach to invention based on a series of approaches that, taken together, aid in thinking about problems in unconventional ways that he identified as being key to most patented inventions (prior to the 1970s). This approach is also known by its acronyms in Russian (TRIZ) and English (TIPS). Altshuller is also known as the "father of TRIZ." Pseudonyms include Genrikh Altov and Henry Altov. Reference [17].

Angel Investor — *See* Seed Capital.

Annual Return — *See* Return on Investment.

Ansoff Matrix — A tool for business growth planning in which Igor Ansoff [18] defined four basic product/market strategies. See Figure 2. In a *Market Penetration* strategy, the idea is to sell more current products into the currently served market or markets (such as by increasing or improving quality, productivity, or marketing). In a *Market Development* strategy, the idea is to sell more current products into a new market or markets (such as with new marketing and sales efforts). In a *Product Development* strategy, the idea is to sell new products into the currently served market or markets, while in a *Product Diversification* strategy, the idea is to sell new products into a new market or markets. The Product Development and Product Diversification strategies usually involve R&D, adopt and adapt, or even purchase or licensing-in of technologies, and then commercialization.

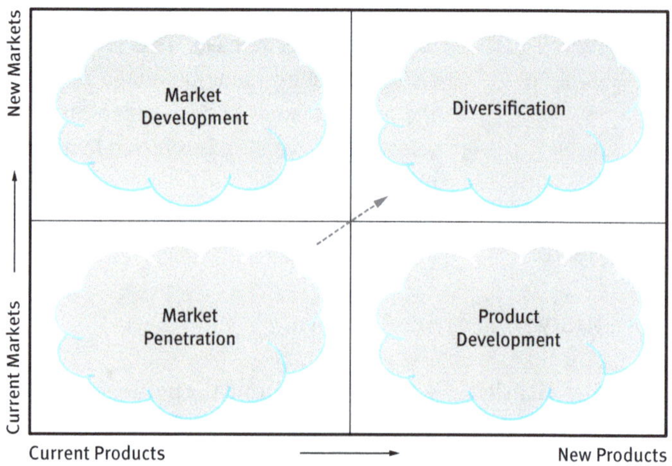

Figure 2: Illustration of Ansoff's Market/Product Matrix. The broken arrow indicates the direction of increasing risk.

A Posteriori — *See A Priori.*

Applied Research — *See* Research and Development.

Applied Research Organization — *See* Research and Technology Organization.

A Priori — A Latin phrase meaning "from the earlier." This phrase is usually used as an adjective to describe a type of knowledge. For example, *a priori* knowledge is knowledge that can be deduced from reason and logic. The converse of *a priori* is *a posteriori*, meaning "from the latter." For example, *a posteriori* knowledge is knowledge that comes only from experience and observation. Much of the knowledge in mathematics is *a priori*, whereas much of the knowledge in the other natural sciences and engineering is *a posteriori*.

Architects — A characterization of one of four kinds of organizational approaches to innovation strategy. "*Architects*" tend to be in mature markets, with intensive capital and resource requirements and with a centralized, structured approach to innovation. They also tend to be focused on customers and competitors for new insights and opportunities, outsourcers of development and prototyping, and innovating through a top-down, formal process. Jaruzelski and Dehoff have referred to such organizations (and also to "Moonlighters") as "*Market Readers*" [19]. *See* Innovation Strategy Mapping. *See also* Top-Down Innovation.

Architectural Innovation — A form of *Evolutionary Innovation* in which improvements are made to the linkage(s) between the components in a product, process, or service, but not to the components themselves. This is in contrast to another form of evolutionary innovation, *Modular Innovation* (also termed *Component Innovation*), in which improvements are made to one of more of the components in a product, process, or service, but not among the linkages between those components. Reference [20]. *See* Figure 3. In some literature, the term architectural innovation is used to mean disruptive innovation, but this is not recommended. *See* Evolutionary Innovation. *See also* Innovation, Disruptive Innovation, Incremental Innovation, Abernathy-Clark Model, Henderson-Clark Model.

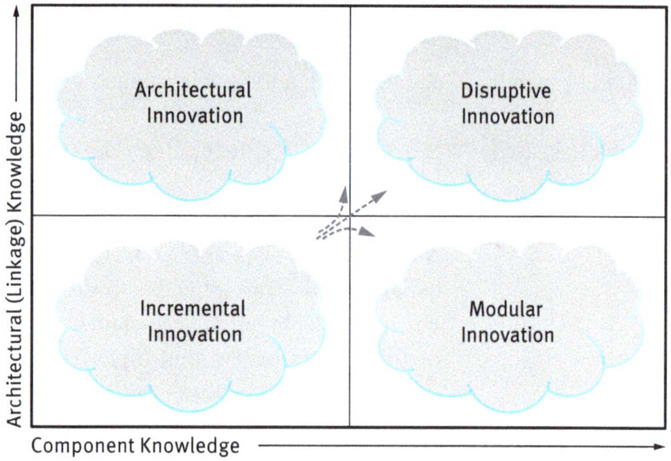

Figure 3: Illustration of Henderson and Clark's Forms of Evolutionary Innovation.

ARIZ — Algorithm of Inventive Problems Solving (ARIZ is the Russian acronym) One of the tools used to try to solve invention problems in Altshuller's Theory of Inventive Problem Solving. *See* Altshuller, Creative Thinking Models, Theory of Inventive Problem Solving.

A-Round Capital — *See* Start-Up Capital.

Art Technology — A technology that is developed or deployed using know-how, craft skill, or experience rather than formal scientific or engineering knowledge, as would be the case with scientific or engineering technologies. Reference [6].

Assets — Owned things having significant value owned by a person or organization, such as "Fixed Assets" such as machinery, large equipment, buildings, and land; "Current Assets" such as cash, short-term securities, accounts receivable, and inventory; and "Intangible Assets" such as patents, copyrights, trademarks, and licenses.

ATAR Model — *See* Awareness, Trial, Availability, Repeat Model.

Augmented Product — *See* Product Platform.

Avoiders

A category of potential customer or technology adopter in a psychological model of technology adoption. *See also* Technology Acceptance, Technology Acceptance Model, Technology Readiness Index, Technology Adoption Lifecycle.

Awareness, Trial, Availability, Repeat Model

(ATAR model) A sales- or profits-forecasting model based on advertising and brand-awareness data. It was developed for consumer product sales, but could be used for process or service sales as well. A version of the model is:

Sales = (purchasing unit #) × (% aware) × (% available) × (% trialing) × (% repeating), and
Profit = number of units sold × (revenue per unit – cost per unit), where

purchasing unit # is the number of purchase participants (people, households, companies, or departments) in a sales region; *percent aware* refers to the fraction of purchasing units that have heard about the new product; *percent available* refers to the probability that the product is available to the purchasing unit; *percent trialing* refers to the fraction of purchasing units that have purchased at least once; and *percent repeating* refers to the fraction of purchasing units that have purchased more than once. Reference [21].

B

Back of the Envelop Calculation	*See* "Does it pencil?"
Barrier to Innovation	*See* Innovation Barrier.
BASIC	Brazil, South Africa, India, and China. *See* BRIC Nations.
Basic Requirements	*See* Global Competitiveness Index.
Basic Research	*See* Research and Development.
Bayh-Dole Effect	A reference to the tendency for university patenting activity to increase following the enactment of legislation enabling universities and/or their researchers to patent inventions arising out of federally funded research programs. The name refers specifically to the U.S. Bayh-Dole Act (1980), but the term has been applied to similar situations in other countries as well. Reference [22].
Benchmarking	The process of comparing an organization's processes and performance measures to the best practices of a specified peer group. Originally a business management tool, benchmarking has also been applied to non-commercial organizations such as not-for-profit organizations (NFPs) including charities, non-government organizations (NGOs), research and technology organizations (RTOs), and even government departments and agencies. In any of these cases, the purposes of benchmarking are usually to assess organizational performance, efficiency, or effectiveness against best practices and/or to improve organizational performance. Also termed Best-in-Class Benchmarking, Best Practice Benchmarking, Process Benchmarking.
BERD	Business Enterprise Expenditure on R&D.

DOI 10.1515/9783110429176-003

Best Available Technology	The best state-of-the-art technology available at a given point in time. In some industries, regulations require purchases to meet a *Best Available Technology* standard. This can be a way to differentiate a product, process, or service in the marketplace. A synonym is Best Practicable Means. *See also* 20/30 Rule. Reference [6].
Best-in-Class Benchmarking	*See* Benchmarking.
Best Practicable Means	*See* Best Available Technology.
Best Practice Benchmarking	*See* Benchmarking.
Beta Release	*See Beta* Test, *Alpha* Test.
Beta Test	A test of a prospective design model product, process, or service conducted prior to commercial production and release [5]. *Beta* testing generally also involves having the users provide feedback on their experiences with the new product, process, or service. *See* Alpha Test. *See also* Prototype.
Beta Version	*See Beta* Test.
BI	Business Intelligence. *See* Competitive Intelligence.
Bibliometrics	*See* Scientometrics.
Big Four Nations	Brazil, Russia, India, and China. *See* BRIC Nations.
Black Box	In the innovation world, the "*black box*" refers to the technological innovation process(es) as a system with unknown components and unknown processes. In some economic models, technological innovation is a black box into which would enter various inputs and out of which would arise outputs in the form of commercializable technologies. Sometimes referred to as the "*Black Box of Innovation.*"
Black Holes	*See* Innovation Performance Mapping.

Black Space — *See* White Space Mapping.

Black Swan Events — *See* Foresight.

Blindside Curve — *See* S-Curve.

Blowback Innovation — *See* Reverse Innovation.

Boilerplate — In the innovation world, boilerplate is any text or code that can be used, and reused in new applications, without needing significant revision. This term is often used with reference to legal agreements, marketing documents, and even computer code (hence "*Boilerplate Code*").

Boom–Bust Cycle — *See* Economic Cycle.

Bootstrapping — *See* Financing.

BOP Innovation — *See* Innovation at the Bottom of the Pyramid.

Bottom-Up Economic Development — *See* Top-Down Economic Development.

Bottom-Up Innovation — *See* Top-Down Innovation.

Brainstorming — *See* Creative Thinking Models.

Brand — In business, a brand comprises the identity of an organization as distinguished from others, particularly from its competitors, and can serve to help distinguish an organization's products, processes, or services from those of its competitors. It can include tangible features, such as a name, logo, tagline, trademark, or other symbol or image. It can also include intangible features, such as how the organization is perceived by others and whether there is an implied promise to stakeholders associated with the brand. The manner in which a brand is perceived by stakeholders is sometimes termed "Brand Reputation." Although a brand is at least partly intangible, it is still given

	accounting treatment as a financial asset and the value of a brand is sometimes termed Brand Equity. *See also* Perceptual Equity, Trademark.
Brand Equity	*See* Brand.
Brand Reputation	*See* Brand.
Breakthrough	A theory, concept, or application that, for the first time (in the world), overcomes a substantial barrier that had been holding back progress among some kind of regional, national, or international peer group. The related term "New-to-the-World Solution" refers to a breakthrough in business and/or innovation. *See also* Camelot Scenario, Innovation Barrier.
BRICK	Brazil, Russia, India, China, and South Korea. *See* BRIC Nations.
BRIC Nations	One of several acronyms that attempt to group together major countries that are in a similar stage of "emerging" economic development. BRIC (or "Big Four") stands for "Brazil, Russia, India, and China," BRICK (or BRICS) stands for "Brazil, Russia, India, China, and South Korea," BASIC stands for "Brazil, South Africa, India, and China."
BRICS	Brazil, Russia, India, China, and South Korea. *See* BRIC Nations.
Bridging Organization	*See* Intermediary Organization.
Broker	*See* Intermediary Organization.
Building Cycle	*See* Kuznets Cycle.
Bush, Vannevar (1890–1974)	An American engineer and science administrator best known for his influence on government's role in the U.S. research, development, and innovation system. In addition to his own research and academic career, he also served on and/or led numerous national-scope research funding and advisory committees culminating in pioneering the role of science adviser to government. He became head of the U.S.'s World

War II National Defense Research Committee (NDRC) and then the Office of Scientific Research and Development (OSRD). These government R&D organizations helped achieve numerous wartime technological innovations in such areas as radar, explosives, and nuclear weapons. Bush also developed the U.S.'s post-war science strategy, which included financial aid for discovery and applied research, and led to the creation of the U.S. National Science Foundation (NSF). References [23, 24].

Business Accelerator — An organization, facility, and/or service that provide modest amounts of capital, mentorship, and/or other services to a business start-up or early-stage business in return for small amounts of equity. The time spent in an accelerator is small, often only a few months. *See also* Business Incubation, Business Incubator.

Business Angel — Angel Investor. *See* Seed Capital.

Business Case — The essential basis for the market potential and/or the commercial viability of a product, process, or service. More detaois regarding the market and the anticipated revenues and expenses would be contained in the Business Model, while details on the whole business process from development through to sales would be contained in the Business Plan. See also Business Concept, Business Model, Business Plan.

Business Concept — An idea for commercializing a new product, process, or service that has been expanded into a description of the nature of the new product, process, or service; what is does; how it is different and better than what is already in the marketplace; how it would be delivered; and who would be the customers for it. Thus, a business concept is often referred to as being "*a bridge between an idea and a business plan.*" *See also* Product Definition, Business Case, Business Plan.

Business Culture Innovation — *See* Administrative Innovation.

Business Cycle — *See* Economic Cycle.

Business Development — The processes of identifying which business areas in a company can and should grow, developing strategies to achieve such growth, and it may also involve implementing such strategies. Business Development builds on the positioning created through Marketing and helps focus Sales. It may also identify opportunities for strategic alliances, mergers and acquisitions or, conversely, opportunities for divestitures. The Business Development function usually also includes the process of identifying and nurturing customer relationships, in partnership with the Sales function. *See also* Marketing, Sales, Distribution.

Business Ecosystem — An organized network of organizations that collectively support a product, process, or service business. Also termed "*Virtual Cluster*." The anchor and/or leader of a business ecosystem is sometimes referred to as the "*Ecosystem Leader*," "*Keystone*," or "*Platform Leader*" [25]. The other ecosystem organizations are sometimes referred to as "*Niche Players*," or "*Complementors*," whose products and services contribute to the productivity and outputs of the greater ecosystem. Examples: Wal-Mart, Apple, and Mozilla have been ecosystem leaders for their respective business ecosystems. A business ecosystem is sometimes referred to as an *Innovation Ecosystem*, but this term is better used to refer to the broader meaning of the important entities involved in technological innovation (i.e. industry, academia, intermediaries, and government). *See also* Innovation Ecosystem.

Business Growth Planning — *See* Ansoff Matrix.

Business Incubation — The provision of services and/or facilities in order to enable and/or accelerate the development and growth of new, usually small- and/or medium-size enterprises (SMEs), from the start-up phase through to some level of stand-alone maturity. The services may span a wide range from standard business services, such as legal, financial, recruitment, office facilities, and marketing, to custom services, such as access to angel- and venture-financiers and successful entrepreneurs, research, development, access to specialized facilities, prototyping, and demonstration. The goals in business incubation are usually to enable SME growth while reducing the associated risks, costs, and/or time involved. *See also* Business Accelerator, Business Incubator, Small- and/or Medium-Sized Enterprise.

Business Incubator — An organization, facility, and/or service in which significant amounts of capital, mentorship, management, education, infrastructure, and/or other services are provided to a business start-up or early-stage business, usually in return for significant amounts of equity. The goal is to provide enough support for entrepreneurs to survive the *valley of death* period between initial financing and significant sales of the new business' products, processes, or services. Depending on the nature of the business, this could take several months to several years. Also termed Technology Hatchery. *See also* Business Incubation, Business Accelerator, Cluster, Research and Technology Park, Valley of Death.

Business Intelligence — (BI) *See* Competitive Intelligence.

Business Logistics Management — *See* Value Chain.

Business Model — A description of the manner in which a business earns revenues and profits. Business models range from very simple, and brief descriptions to more complex, and lengthy descriptions resembling the business plans. *See also* Business Case, Business Concept, Business Plan, Commercialization Plan.

Business Model Innovation — *See* Strategic Innovation.

Business Model Trap — *See* Innovation Barrier.

Business Plan — A plan that describes the intended process for taking a new product, process, or service concept through the process of development, commercialization, and into an actual business enterprise. The business plan often evolves out of a *Commercialization Plan* and will typically cover descriptions of the overall project and value proposition; the new product, process, or service and its current development status; the *Market Analysis* and a marketing strategy; the *Intellectual Property*; an operations plan; a management plan, including descriptions of the key personnel; financial information and projections; the risk analysis, including how the principal risks are being managed; and any other special circumstances. *See also* Business Case, Business Concept, Business Model, Commercialization Plan.

Business Plan Innovation — *See* Strategic Innovation.

Business Process Innovation — *See* Administrative Innovation.

Business R&D Intensity — (Innovation Indicator) An indicator of the R&D investments by and within the business sector is the ratio of business-enterprise funded R&D to gross domestic product (BERD/GDP). *See* Innovation Indicators and Tables 4 and 8.

Table 4: Examples of Innovation Indicators for Regions or Countries. References [85, 86]

Indicator	Explanation	Background
R&D Intensity	The ratio of gross domestic expenditure on R&D to gross domestic product (GERD/GDP).	A broad indicator of R&D investments from all sectors in an economy.
Business R&D Intensity	The ratio of business enterprise-funded R&D to gross domestic product (BERD/GDP).	An indicator of R&D investments by and within the business sector.
Government R&D Intensity	The ratio of government budget appropriations or outlays for R&D to gross domestic product (GBAORD/GDP).	An indicator of R&D investments by governments.
Investment in Knowledge	The ratio of total knowledge investments to gross domestic product.	"Knowledge investments" include: higher education, R&D, and software
ICT Investment Intensity	The ratio of ICT expenditure to gross domestic product.	Expenditures in the information and communication technology (ICT) sector.
R&D Personnel	The number of researchers involved in R&D per 10,000 personnel in the labor force.	An indicator of the total number of people directly involved in R&D, in an economy.
Patent Applications	The number of patent applications* per 10,000 personnel in the labor force.	An indicator of the total number of practical inventions being protected in an economy.
External Patent Applications	The number of external patent applications per 10,000 personnel in the labor force.	An indicator of the total number of internationally relevant practical inventions being protected.

Table 4 (continued)

Indicator	Explanation	Background
Triadic Patents Issued	Patents in triadic patent families issued per 1 million people in the population.	Triadic patent families are groups of patents that have been granted in multiple (3 or more) countries on the same invention.
Percentage of patents with foreign co-inventors	100 times the ratio of the number of patents having foreign co-inventors to the total number of patents issued in a country	An indicator of the degree of connectedness and collaboration among a country's inventors and those in other countries.
Trademarks	The number of trademarks registered per 10,000 personnel in the labor force.	An indicator of the total number of new products and services being protected in an economy.
Connectedness Index (ICT)	A blended indicator calculated based on such product and service factors as the market-ready supply of infrastructure, networks, and systems, demand, price, and usage.	Example: The Conference Board of Canada Connectedness Index.
Ease of Entrepreneurship Index	A blended indicator calculated based on such factors as barriers to competition, regulatory and administrative opacity, and administrative burdens.	Example: The Conference Board of Canada Ease of Entrepreneurship Index.
Venture Capital Intensity	The ratio of venture capital investments to gross domestic product.	
Technology Balance of Payments	The net technology transactions (purchasing power parity USD) per 10,000 personnel in the labor force.	The balance of sales versus purchases of technology (such as patents, licences, designs, trademarks, and trade secrets) in an economy.

* This can be taken to be either the national patent applications (meaning by both residents and non-residents) or the resident (only) patent applications. The former includes an indication of potential technology diffusion to come from other countries).
** Here, "significant" means inventions of such impact that they are worth protecting in other countries beyond that in which the inventions were made.

Business Structure Innovation *See* Administrative Innovation.

Buying Hierarchy A new product life-cycle concept attributed to Windermere Associates (San Francisco, CA, USA). In this model, a new product goes through an evolutionary sequence comprising four phases: (1) functionality, in which the new product is able to compete based on a functionality not satisfied by competitors; (2) reliability, in which the new product is able to stay ahead of competitors based on market demand for reliability; (3) convenience, in which the new product competes by providing functionality, reliability, and superior convenience; (4) price, in which the earlier attributes are all satisfied by competitors in the marketplace and competition comes down to price. *See also* Product Life-Cycle Curve in the entry for S-Curve.

Buying Unit — *See* Awareness, Trial, Availability, Repeat Model.

C

Ca.	See *Circa.*
Camelot Scenario	An imaginary scenario in which some kind of substantial barrier that had been holding back progress in business and/or innovation does not exist. The creation of Camelot Scenarios is a brainstorming tool that is sometimes used in an attempt to find a solution or breakthrough. *See also* Breakthrough, Innovation Barrier. Reference [9].
CapEx	*See* Capital Expense.
Capital	The assets, such as cash, investments, and property that a person or organization has available. In the context of innovation, this refers to the financial resources to support the research, development, and commercialization of a new product, process, or service. Debt capital refers to some form interest-bearing loan, often from a financial institution. Grant capital refers to some form of (usually) government grant assistance. Equity capital refers to some form of partial ownership in return for financial investment. Other examples of capital include Sweat Equity, Seed Capital, Pre-Venture Capital, and Venture Capital.
Capital Expense	(CapEx) A financial term for an expenditure for something that is intended to provide benefit to an organization for more than a year, as opposed to an "Operating Expense" (OpEx), which refers to expenditures for things that are used or done within an operating year. Accordingly, capital expenses are recognized on financial statements over time through annual depreciation amounts, while operating expenses are recognized within the year they are incurred. CapEx examples include major equipment and facilities. OpEx examples include research and development (R&D) and selling, general, and administrative (SG&A) expenditures.
Captive Inventor	An inventor who works under a contractual arrangement that assigns intellectual property ownership to someone else, usually their employer.

DOI 10.1515/9783110429176-004

Carve-Out — A synonym for Spin-Off. *See* Spin-Out.

Cash Flow — The difference between the amount of money flowing in during a specified period of time (often a month) and the amount flowing out during the same period. If the inflow of money is greater, it is referred to as *positive cash flow*; the converse is referred to as *negative cash flow*.

Catalytic Innovation — *See* Disruptive Innovation.

Caveat Innovator — A term coined by Richard Foster with reference to the perils of ignoring the implications of ignoring the plateauing nature of S-curves. According to Foster, the evolutionary approach "is doomed to fail" in the face of discontinuities brought about by competing technologies or product/services. Reference [26]. *See* Foster, S-Curve.

CDA — Confidential Disclosure Agreement. *See* Non-Disclosure Agreement.

CE — *See* Concurrent Engineering.

Central Chain-of-Innovation — *See* Chain-Linked Model.

Chain-Linked Model — An example of a nonlinear "*coupling model*" of the technological innovation process, which was proposed by Stephen Kline in 1985. In the Chain-Linked Model, the innovation process begins with the identification of a potential market and then proceeds through the conception of a design (possibly but not necessarily involving research), development of the design, prototyping and testing, then redesigning and retesting, then production, marketing and distribution. This principal pathway has been termed the "central chain-of-innovation." Supplementing and interconnecting with this principal pathway are a number of feedback loops including loops that feed-in knowledge and research results of various kinds such that there is not a single pathway but at least five major possible pathways from beginning to end (see in Figure 4). In the chain-linked model, research processes, for example, can feed most of the other elements in the process. Also termed Linked-Chain Model, Chain of Innovation. References [27, 28]. *See also* Non-Linear Innovation Models.

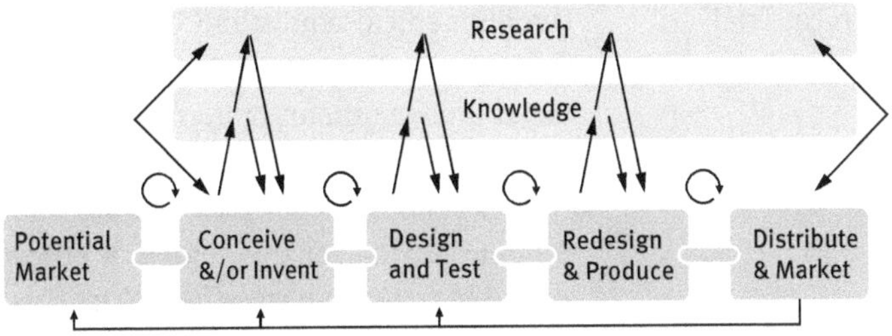

Figure 4: Illustration of the "Chain-Linked Model" of innovation, in which feedback loops and alternative developmental pathways are superimposed on a central chain-of-innovation.

Chain of Innovation	*See* Chain-Linked Model.
Chasm	*See* Tipping Point.
Chesbrough, Henry W. (1956 – Present)	An American business professor known to the innovation world for his 2003 book on *Open Innovation*. Although the activities now associated with open innovation were already being practiced, especially since the 1980s, Chesbrough's work helped coalesce them into an overall approach that could be incorporated into an organization's strategy. He also connected such processes as the acquisition of external technologies and other information (the inbound open innovation process) and the use of internal technologies and other information externally (the outbound open innovation process). References [29, 30].
Chief Innovation Officer	*See* Participative Innovation.
Christensen, Clayton M. (1952 – Present)	An American business professor best known to the innovation world for his works on business strategy and technological innovation by businesses. He coined the phrases "Innovator's Dilemma" and "Technology Mudslide Hypothesis," among others. Reference [14].
CI	*See* Competitive Intelligence.
Circa	A Latin term meaning "*around*" or "*about*." It is frequently used to mean "*approximately*," particularly when referring to historical dates. There are many abbreviations for the term *circa*, including *c.* and *ca*.

Clinical Trial The biomedical equivalent of a pilot or field test, in which a new, prospective medical or veterinary medical treatment is tested on live human or animal subjects. There are four standard phases of clinical trial: Phase 1 refers to testing for safety, Phase 2 refers to testing for the possibility of efficacy, Phase 3 refers to testing for the degree of efficacy, and Phase 4 refers to testing for long-term effects. The first three phases are generally considered to be part of the research and development (R&D) process, while the fourth phase is usually only considered to be part of R&D if it results in an additional scientific or technological advance. Reference [7]. *See also* Field Pilot.

Closed Innovation *See* Open Innovation.

Closed Science *See* Open Science.

Clust-Bun Effect *See* Clustering-Bunching Effect.

Cluster (Innovation Cluster) A group of organizations in fairly close proximity comprising companies with similar business interests and often supporting or coordinating organizations. These organizations normally share sufficient common interests that they become interconnected in formal and/or informal ways. The cluster theory of economic development holds that such clusters strengthen both entrepreneurship and innovation by providing a supportive, knowledge-rich, and resource-rich environment. For the same reasons, a cluster environment can be particularly attractive to start-up and small- and medium-sized enterprises. Clusters are normally led by industry but are frequently supported by government agencies and/or government funding. Also termed Innovation Cluster, Cluster Ecosystem, or Regional Innovation Cluster. *See also* Business Incubator, Innovative Regional Cluster.

(Technology Cluster) A technology cluster is a grouping of two or more technologies that are closely related, especially as perceived by customers. Example: a household paper recycling service and a bottle/can recycling service. Rogers found that some businesses market clusters of technological innovations because they find more rapid consumer acceptance that way [31].

Cluster Ecosystem	*See* Cluster (Innovation Cluster).
Clustered Regularly Interspaced Short Palindromic Repeats	*See* CRISPR.
Clustering-Bunching Effect	A reference to the tendency for investment funds to flow to locations that have experienced bursts and/or clusters of inventions and/or technological innovations. Thus, the *clustering* of the inventions/innovations is viewed as leading to the *bunching* of investment funds, which intensify the business activities. Also termed the Clust-Bun Effect. Reference [32].
Codified Knowledge	*See* Knowledge.
Cognitive Innovation	The ideation, development, and deployment of improvements in ways of thinking and conceptual models. Example: lateral thinking. This is a form of non-commercial innovation. *See* Innovation (Non-Commercial). *See also* Creative Thinking Models, Social Technology. Reference [33].
Cognitive Tacit Knowledge	*See* Social Technology.
Collaborative Compromise Paradox	*See* Innovation Paradoxes (Innovation Process Paradoxes).
Collaborative Innovation	A form of open innovation in which two or more organizations work together on the development of innovation(s). Such organizations can include multiple customers, suppliers, and technology developer/providers. Advantages can include efficiency, and cost and risk sharing, plus the ability to leverage each other's technological and inventive capacities. Disadvantages can include difficulties maintain cohesion and focus, and/or intellectual property issues. This is sometimes termed Participative Innovation, although the latter term can have other meanings as well. *See also* Innovation Ecosystem, Participative Innovation, Open Innovation.

Collectors	*See* Early Majority.
Comets	*See* Innovation Performance Mapping.
Commercial Innovation	The meaning of innovation as originally defined by Schumpeter [3, 4]. *See* Innovation. The term *Commercial Innovation* issometimes used to distinguish from *Non-Commercial Innovation*, which refers to any aspects of improving an organization's efficiency or effectiveness that manifest themselves in ways other than the introduction into the marketplace of commercial products, process, or services. Example: In some usage, the term *Commercial Innovation* is used to specifically exclude military innovations. *See also* Innovation (20th–21st century).
Commercialization	The process of developing a new product, process, or service into a form that is market-ready and introducing it into the marketplace (product launch). The commercialization process generally follows the discovery and invention processes, if these are needed. *See also* Business Planning, Commercialization Plan, Market Analysis, Product/Process/Service Development, Technology.
Commercialization Gap	*See* Innovation Gap.
Commercialization Gap Theory	*See* Innovation Gap Theory.
Commercialization Plan	A plan that describes the intended process for taking a new product, process, or service concept through the process of research, development, demonstration, and commercialization. A commercialization plan will typically cover descriptions of the overall project and value proposition; the new product, process, or service and its current development status; the *Market Analysis*; the *Intellectual Property*; the risk analysis, including how the principal risks are being managed; and the next steps in the Product/Process/Service Development Plan. A commercialization plan is often later expanded into a *Business Plan*. *See also* Business Plan.
Commercialization Valley of Death	*See* Valley of Death.

Table 5: A Generalized Description of Commercial Readiness Index (CRI) Levels*

Commercial Readiness Level	Description
CRI 1	The technology is commercially prospective but has not yet been commercially tested or proven.
CRI 2	The technology has passed a first, small-scale, commercial trial.
CRI 3	The technology has been commercially deployed, driven at least partly by market-pull.
CRI 4	Multiple commercial deployments have occurred, possibly with some level of government support.
CRI 5	Market-pull is driving broad commercial deployment and competition is emerging in the marketplace.
CRI 6	Mature commercial deployment with established standards, regulatory acceptance, and performance track-record.

* Based on Reference [34].

Commercial Readiness Index — (CRI) A representation of the readiness of a mature technology, and an extension of the Technology Readiness Level (TRL) scale, transcending TRL 9. The original Australian CRI system uses a scale ranging from 1 to 6 (*See* Table 5 and Figure 5). Reference [34]. *See also* Technology Readiness, Technology Readiness Level.

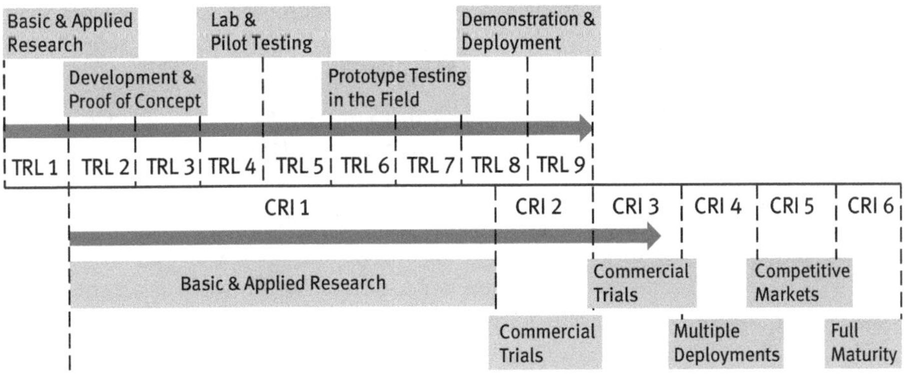

Figure 5: Illustration of Technology Readiness Levels and Commercial Readiness Index Levels.

Commitment Trap — *See* Innovation Barrier.

Communication Model — A term referring to, or a version of, the "*coupling model*" of the technological innovation process. *See* Non-Linear Innovation Models.

Compartmentalization	An approach to developing a complex device, or process, in which individual components are developed and tested before integrating them into the overall system [5].
Competence-Destroying Innovation	*See* Competence-Enhancing Innovation.
Competence-Enhancing Innovation	An innovation process that adds to and reinforces competencies, as opposed to "*Competence-Destroying Innovation,*" which supersedes previously existing competencies and may even make the latter obsolete. For example, incremental innovation is competence-enhancing, whereas disruptive innovation is usually competence-destroying. An example of competence-enhancing innovation is the electric typewriter over the mechanical typewriter, while an example of competence-destroying innovation is the transistor over the vacuum tube. Reference [35].
Competitive Intelligence (CI)	Knowledge about developments and trends relevant to an organization's competitive position. Such knowledge could relate to discoveries, inventions, and/or technologies; products, processes, and/or services; partners, competitors, and/or suppliers; and customers and markets, for example. Competitive intelligence requires human analysis and judgement, and encompasses data, information, and knowledge (see also *Wisdom Hierarchy*). Competitive intelligence may refer to the process of producing such intelligence, or to the results themselves. Also termed *Business Intelligence*. *See also* Social, Technological, Economic, Environmental, and Political Analysis.
Competitiveness	(Economics) The ability of an organization, country, or region to offer and supply products, processes, services, or jobs that effectively compete with others in the same market(s). The term has also been applied to specific regions or marketplaces, where it is used to refer to the extent to which such regions or marketplaces comprise competitive environments. *See also* Competitors, Macroeconomics, Microeconomics.
Competitiveness Drivers	In making comparisons among countries' global competitiveness, the World Economic Forum (WEF) considers each county's stage of economic development maturity, using three broad categories of competitiveness driver and different weightings

for calculating the Global Competitiveness Index (GCI) within each category. In the WEF categorization, *Stage 1*, or *Factor-Driven Economy*, refers to countries having GDP *per capita* of less than $2,000; *Stage 2*, or *Efficiency-Driven Economy*, refers to GDP *per capita* within the range $3,000 to $8,999; and *Stage 3*, or *Innovation-Driven Economy*, refers to GDP *per capita* of more than $17,000. *Economies in Transition* lie between Stages 1 and 2 or between Stages 2 and 3. Depending on the category, the weightings are adjusted in an attempt to make the GCI appropriate to a given country's degree of economic development and therefore the manner in which it competes globally. For example, a factor-driven economy competes mostly based on the strength of their "basic requirements" such as natural resources, infrastructure, and/or labor; an efficiency-driven economy competes mostly on the strength of their "efficiency enhancers" such as higher education and training, technological readiness, and market size; an innovation-driven economy competes mostly based on the strength of their "innovation and sophistication factors" such as capacity to innovate, value-added manufacturing, and value chains. Reference [36]. *See also* Global Competitiveness Index.

Competitor Intelligence — *See* Competitive Intelligence.

Competitors — A description of the competitors in the marketplace for a product, process, or service, and also how the product is or will be differentiated from them. There are almost always competitors. If there are truly no competitors then there is a good chance there is no market. *See also* Competitiveness, Market Analysis. Reference [6].

Complementor — *See* Business Ecosystem.

Component Innovation — A synonym for "*Modular Innovation.*" *See* Architectural Innovation

Concept-Push Innovation — *See* Design-Driven Innovation.

Concurrent Engineering — (CE) A product development approach in which multidisciplinary functions work together through the entire process. *See also* Design for Excellence.

Confidential Disclosure Agreement (CDA) *See* Non-Disclosure Agreement.

Confidentiality Agreement *See* Non-Disclosure Agreement.

Connectedness (ICT) The availability and use of information and communications technology (ICT) to enable communications, information flows, and trade.

Connectedness Index (ICT) In the area of information and communications technology (ICT), the "Connectedness Index" is a blended indicator calculated based on such product and service factors as the market-ready supply of infrastructure, networks, and systems, demand, price, and usage. Example: The Conference Board of Canada Connectedness Index. *See* Innovation Indicators and Table 4.

(Global Business) In the area of international business and trade, a Global Connectedness Index is a blended indicator calculated based on such cross-border flows as trade, capital, information, and people. It is intended to indicate the breadth and depth of a country's interactions that cross its national borders. Example: DHL Global Connectedness Index (GCI), McKinsey Global Institute (MGI) Connectedness Index.

Conservative *See* Late Majority.

Consumer-Innovators (Innovators) Consumers that make inventions or modify existing products to create the first concepts or even the first prototypes of new technological innovations, especially in the area of products. Example: The skateboard was reportedly first developed and built by children for their own use by hammering roller skate wheel assemblies onto wooden boards. This is sometimes referred to as "User Innovation," especially in the context of user modifications of current commercial products. Reference [37].

(Early Adopters) The early adopters in the technological diffusion process, as described by the Rogers diffusion of innovation model. *See* Figure 6. These consumers are not actually innovators in the sense of making technological innovation happen, however.

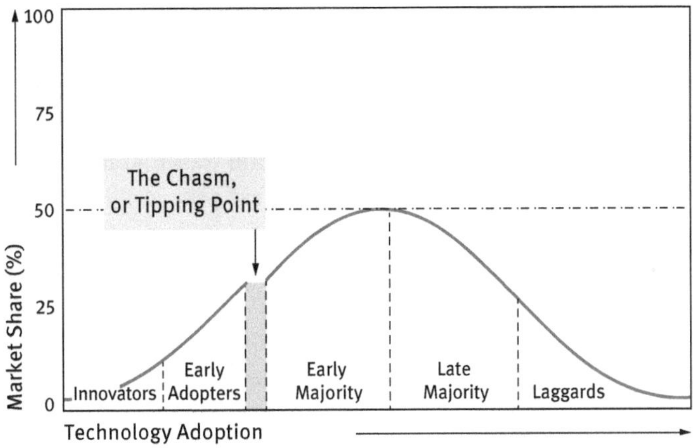

Figure 6: Rogers' Technology Diffusion Model Modified to Illustrate The "*Chasm*," or Tipping Point.

Context-Driven Research — *See* Modes of Science.

Continuous Innovation — *See* Incremental Innovation.

Convergent Thinking — Part of Creative Problem-Solving. *See* Creative Thinking Models.

Cooper, Robert G. — An American management scholar known for his contributions in product development and innovation management. He pioneered many groundbreaking discoveries in product innovation, including the Stage-Gate® Idea-to-Launch Process, which has been widely implemented in North American companies. References [38,39].

Copyright — A legal protection for ownership of the original author(s) of published or unpublished artistic or literary creations, including computer software. The specific rights and terms are determined by a given country's copyright law. *See also* Intellectual Property.

Core Innovation — *See* Incremental Innovation.

Corporate Immune System — A corporate cultural trait in which the internal "system" reacts against anything that is perceived to be a threat to the continuance of current norms and practises and/or a threat to the overall system. Although such a culture can help protect against things that could harm the organization, it can also repel new ideas and practises that could benefit the organization.

Corporate Myopia	A corporate cultural trait that focuses on current “day-to-day” status quo business practises to the exclusion of potential or even prospective future business practises (hence the analogy with nearsightedness). Although such a culture can help with maintaining focus on the things that currently provide business value, it can also repel new ideas and practises that could benefit or even be critical to the organization in the future.
Cost Innovation	*See* Reverse Innovation.
Cost Minimization	*See* Late to Market.
Country-First	*See* World-First.
Coupling Model of Innovation	*See* Non-Linear Innovation Models, Chain-Linked Model. *See also* Generations of Innovation.
CPS	Creative Problem-Solving. *See* Creative Thinking Models.
Creative Accumulation	*See* Schumpeter Mark I Innovation.
Creative Age or Society	*See* Technological Ages.
Creative Agglomeration	*See* Schumpeter Mark I Innovation.
Creative Problem-Solving	*See* Creative Thinking Models.
Creative Thinking Models	A number of different approaches to creative thinking, creative problem-solving, and/or inventiveness have been developed, particularly since the 1940s. Listed below are some of the principal foundational approaches. *Brainstorming* stands for “*storming a problem in a commando fashion*” and was coined by Alex Osborn. Osborn’s

brainstorming concept was aimed at groups and was designed to help get lots of ideas out into the open, avoid killing ideas with early criticism, and enable multiple ideas to be combined and/or lead to new ones. Also termed *Think Up* or *Horizontal Thinking*. References [40, 41, 42].

Creative Problem-Solving (CPS) originated with Alex Osborn and Sidney Parnes and is famous for its use of "*divergent thinking*" and "*convergent thinking.*" Divergent thinking is open-ended, wide-ranging thinking used to create a broad set of options. Convergent thinking is solution-oriented thinking used to formulate a specific solution or approach to a problem using the results of divergent thinking. Also termed *Osborn-Parnes Model*. References [41, 42, 43].

Parallel Thinking and *Lateral Thinking* originated with Edward DeBono as creative counters to "*linear*" or "*vertical*" thinking in problem solving. *Parallel Thinking* has to do with avoiding adversarial approaches in team-based creative thinking, focusing instead on more co-operative and constructive approaches. Parallel in this sense means having everyone on the team thinking in, broadly, the same direction. *Lateral Thinking* refers to avoiding thinking about a problem "head-on" and/or in logical "step-by-step" fashion, and instead thinking about it in indirect and/or nonlinear ways (sometimes referred to as *Thinking Out-of-the-Box*). References [44, 45].

Synectics was developed by William J.J. Gordon. The name comes from the Greek word *synektiktein*, referring to the joining together of different, and possibly seemingly irrelevant, ideas in the process of creative problem solving by a group of people. Synectics originally referred to the study of problem-solving and invention by groups, but later evolved into a set of techniques that include what are now known as *Creative Problem-Solving*, *Brainstorming*, and *Lateral Thinking*. Synectics is also a trademarked name. *See also* Medici Effect. References [46, 47].

Theory of Inventive Problem Solving (TIPS, or TRIZ, the Russian acronym) was developed by Genrich Altshuller. TIPS is a systematic approach to invention based on a comprehensive analysis of the patented solutions to hundreds of thousands of previously solved inventive problems. Altshuller identified a series of approaches that, taken together, frequently lead to finding an inventive solution. Reference [17].

Spider Diagrams are diagrams that use text and drawings and/or pictures to visually organize information. Spider Diagrams, or at least their forerunner seems to have originated with the famous notebooks of Leonardo DaVinci. There are other variations of spider diagrams, including *Idea Sun Bursts* and *Mind Maps*.

Creator — *See* Early Adopter and Innovator.

CRI — *See* Commercial Readiness Index.

CRI 1 — Commercial Readiness Index 1. There can be somewhat different definitions, but on a six-point CRI scale, CRI 1 generally refers to the stage at which the technology is commercially prospective but has not yet been commercially tested or proven. *See* Table 5 and Commercial Readiness Index.

CRI 2 — Commercial Readiness Index 2. There can be somewhat different definitions, but on a six-point CRI scale, CRI 2 generally refers to the stage at which the technology has passed a first, small-scale, commercial trial. *See* Table 5 and Commercial Readiness Index.

CRI 3 — Commercial Readiness Index 3. There can be somewhat different definitions, but on a six-point CRI scale, CRI 3 generally refers to the stage at which the technology has been commercially deployed, driven at least partly by market-pull. *See* Table 5 and Commercial Readiness Index.

CRI 4 — Commercial Readiness Index 4. There can be somewhat different definitions, but on a six-point CRI scale, CRI 4 generally refers to the stage at which multiple commercial deployments have occurred, possibly with some level of government support. *See* Table 5 and Commercial Readiness Index.

CRI 5 — Commercial Readiness Index 5. There can be somewhat different definitions, but on a six-point CRI scale, CRI 5 generally refers to the stage at which market-pull is driving broad commercial deployment and competition is emerging in the marketplace. *See* Table 5 and Commercial Readiness Index.

CRI 6 — Commercial Readiness Index 6. There can be somewhat different definitions, but on a six-point CRI scale, CRI 6 generally refers to the stage at which mature commercial deployment with established standards, regulatory acceptance, and performance track-record. *See* Table 5 and Commercial Readiness Index.

CRISPR — (Clustered Regularly Interspaced Short Palindromic Repeats) A new (as of 2013) genome editing tool that enables accurate and rapid targeting, investigation, and splicing and editing of specific DNA sequences in a genome. Potential applications range from drug development, to disease treatment, to genetic trait modifications that could extent to food crops, animals, and humans. Also termed "CRISPR-Cas9," for the first of several nucleases discovered. References [48, 49].

CRISPR–Cas9 — *See* CRISPR.

Critic — *See* Early Majority.

Critical Technology Events — (CTEs) Key advances in knowledge during the development and commercialization of a technology. CTEs have been defined as "ideas, concepts, models, and analyses that had a major impact on the development" of a particular innovation, and which led to significant improvements over the preceding technology [50].

Cross-License — A mechanism for exchanging technologies by which people or organizations grant royalty-free licenses to other people or organizations, in exchange for reciprocal rights to the latter's technologies. *See also* License, Licensing, Royalty-Free License.

CTEs — *See* Critical Technology Events.

Cultification Paradox — *See* Innovation Paradoxes (Innovation Process Paradoxes).

Cumulative Synthesis Model	A model for the innovation process proposed by Abbott Payson Usher, by which a sequence of steps describes the process of creating something novel, practical, and useful (technological innovation). One formulation of these steps is: "*perception of the problem*," followed by "*setting the stage*" (i.e., acquiring the elements necessary for the solution), followed by the "*primary act of insight*" (in which the solution to the problem is found, followed by "*critical revision and development*" (in which the solution is made practical). Usher's model is a hybrid of the transcendentalist and mechanistic models in that somewhere in the process there is still a critical act of inspiration and/or insight. References [51, 52]. *See also* Transcendentalist Model, Mechanistic Model.
Current Assets	*See* Assets.
Customer Intelligence	*See* Competitive Intelligence.
Customer-Oriented Innovation	This refers to maintaining a customer-orientation throughout most or all of the technological innovation processes. Also termed *Outcome-Driven Innovation*. Reference [33].
Customers	A description of the target purchasers, and also of the ultimate end-users, of a product, process, or service. *See also* Market Analysis. Reference [6].
Customer Value-Chain Analysis	(CVCA) An approach to product development in which, beginning with the product-definition phase, design teams identify key stakeholders (including customers, suppliers, etc., all of whom are referred to as "customers" in this case), their interrelationships if any, and their role(s) in the new product's life cycle. CVCA helps design teams to recognize diverse product requirements at an early stage of the development process. CVCA is an example of a Design for Excellence (DFX) tool. Reference [53]. *See also* Design for Excellence, Voice of the Customer.
CVCA	*See* Customer Value-Chain Analysis.
Cycle Time	*See* Time to Market.

D

Dark Factory	An entirely automated factory would have so few people that the lights could mostly be left switched off, hence a "dark factory."
Data Hierarchy	*See* Wisdom Hierarchy.
Data-Information-Knowledge-Wisdom Hierarchy	(DIKW Hierarchy) *See* Wisdom Hierarchy.
Davidow's Law	Essentially a strategy that a company that is the first to release a product may be able to dominate the marketplace, especially if it is the first to obsolesce its own, previously dominating, product. Named for William H. Davidow, an executive at Intel in the 1970s and 1980s. Reference [54]. *See also* The Borg Law.
Death Valley	*See* Valley of Death.
Death Valley Curve	*See* Valley of Death.
Debt Capital	*See* Capital.
Debt Financing	*See* Financing.
Decision-Making Errors	Among the kinds of errors that can occur in a decision-making process are (i) judging that there is a risk, benefit, or opportunity when there is not (*Type I* or *False-Positive Error*) and (ii) judging that there is no risk, benefit, or opportunity, or failing to notice the same, when there is (*Type II* or *False-Negative Error*). In the innovation realm, a false-negative, or Type II, error could involve being too skeptical or too negative about the prospects for a new technology or its market opportunity. *See also* Technology Readiness, Technology Acceptance Model.
Deep Learning	A form of machine learning, in which computers (including computers in robots) use massive data sets and neural

DOI 10.1515/9783110429176-005

networks to learn through experience. This represents a form of artificial intelligence. Also termed Deep Structured Learning, Hierarchical Learning, Deep Machine Learning. *See also* Singularity.

Delphi Method

A foresight tool by which a group of experts in a field are surveyed for their opinions about a possible future situation or scenario, a summary of the entire group's responses and rationales is provided as feedback, and then the group is surveyed again one or more times until something approaching a consensus, the "*expert consensus*," is achieved. Originally developed by the RAND Corp. (Santa Monica, CA, USA) in the 1950s with military foresight in mind, the Delphi Method has since been widely used in foresight activities of all kinds. *See also* SWOT Analysis, STEEPV Analysis.

Delta

(Business Planning) The difference between the revenues that an organization is forecasting to generate and the revenues that they have targeted to generate.

Demand-Induced Innovation

See Linear Innovation Models.

Demand-Pull Innovation

See Linear Innovation Models.

Demographic Cycle

See Kuznets Cycle.

Demonstration

(1) In some usage, this is simply a physical demonstration to show that an invention basically works. Such demonstration could involve an engineering prototype, or at least a working model. Also termed Proof of Concept. *See also* Prototype.

(2) In other usage, this is a demonstration and/or test of a production prototype of a new or improved product, process, or service conducted at full-scale and under conditions of actual field, industrial plant, or market operation. With this definition, "demonstration" is usually the next step in commercialization beyond the engineering prototype and/or field test step. Sometimes termed *Field Demonstration, Plant Demonstration,* or *Market Demonstration. See* Figure 7. *See also* Field Pilot, Production Prototype, Prototype.

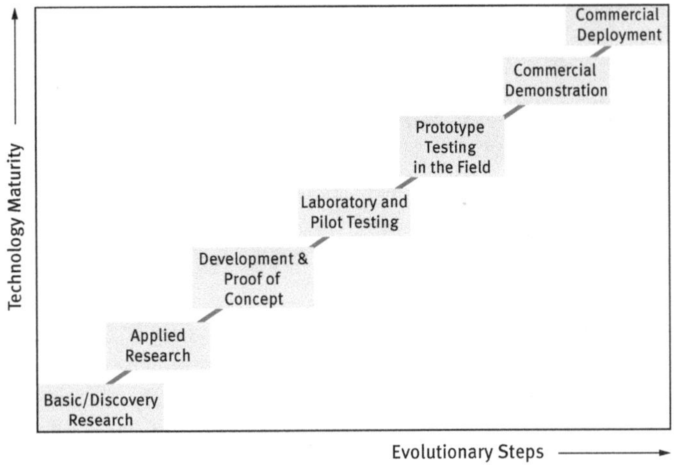

Figure 7: A Simple Technology-Push, Linear Model of Technological Innovation.

Deployment *See* Technology Deployment.

Design-Driven Innovation A form of technological innovation that is not "market pull" and also not quite "technology push" but rather "*concept push*" – meaning that the innovation relates to an entirely new product or service concept that was not previously imagined and which prospective customers did not previously know that they might want and/or need. Where they are successful, design-driven innovations are frequently breakthrough innovations. Also termed Concept-Push Innovation. Examples include the Wii entertainment system, which combined video gaming with physical activity and social interaction, and the Swatch, which transformed timepieces into fashion accessories. Reference [55]. *See also* Disruptive Innovation.

Design for All *See* Design for Excellence.

Design for Assembly (DFA) *See* Design for Excellence.

Design for Excellence (DFX) A suite of product development design programs in which customer needs are identified at an early stage in the process. Some specific DFX tools include "*Design for Assembly*" (DFA), "*Design for Manufacture*" (DFM), "*Design for Manufacture and Assembly*" (DFMA), "*Design for Excellence*" (DFX), and "*Design*

	for All." References [56, 57]. *See also* Concurrent Engineering, Customer Value-Chain Analysis.
Design for Manufacture	(DFM) *See* Design for Excellence.
Design for Manufacture and Assembly	(DFMA) *See* Design for Excellence.
Design for Reuse	The including of previously designed components in a new product, process, or service. The term originated with reference to the use of previously designed code or data in the software engineering field, but is now used more broadly. Also termed Design Reuse. Reference [58].
Design for X	(DFX) *See* Design for Excellence.
Design Model	*See* Production Prototype.
Design Patent	*See* Patent.
Design Reuse	*See* Design for Reuse.
Development	Experimental Development. *See* Research and Development.
DFA	Design for Assembly. *See* Design for Excellence.
DFM	Design for Manufacture. *See* Design for Excellence.
DFMA	Design for Manufacture and Assembly. *See* Design for Excellence.
DFX	*See* Design for Excellence.
DHL GCI	DHL Global Connectedness Index. *See* Connectedness Index.
DHL Global Connectedness Index	(GCI) *See* Connectedness Index.
Diffusion	*See* Technology Dissemination.

Diffusion of Innovations Model — *See* Rogers.

Diffusion of Technologies Model — *See* Rogers.

Digital Age — *See* Technological Ages.

DIKW Hierarchy — Data-Information-Knowledge-Wisdom Hierarchy. *See* Wisdom Hierarchy.

Direct Indicators — *See* Innovation Indicators.

Dirigiste Regional Innovation System — *See* Regional Innovation System Types, Table 6.

Table 6: A Simplified Taxonomy for Regional Innovation Systems.

Regional Innovation System Type	**Technology Transfer Initiation**	**Funding**	**R&D Competence**	**Technical Specialization**	**Degree of System Coordination**
Grassroots	Local	Diffuse, mostly local	Mostly very applied R&D	Low, geared to problem-solving	Low
Network	Multi-level: local, regional, external	Guided by companies, banks, government	Discovery research plus applied R&D aligned with companies	Flexible, wide ranging	High, with many stakeholders
Dirigiste	External to the region	Centrally determined by government	Mostly discovery research	High	High, by government

Regional Innovation System Type	**Innovation Drivers**	**R&D Depth**	**Degree of Associations**
Localist	Mostly driven by small companies	Limited, few public R&D resources	High, involving entrepreneurs and regional government
Interactive	Balance of drivers from large and small companies	Mix of public and large company resources	High, involving regional industry networks and associations
Globalized	Dominated by global companies	Determined by large company resources	Low, unless led (and directed) by large companies

Discontinuity	*See* S-Curve.
Discontinuous Innovation	*See* Disruptive Innovation.
Discoveror	A person that has discovered or co-discovered created a new addition to the world's body of knowledge. Discoveries are generally attributed to the first to document, publicly present, or publish their discoveries. *See also* Discovery, Innovator, Inventor, Patent.
Discovery	A new addition to the world's body of knowledge. Such an addition could be developed through the formulation and testing of new theories or through the observation and study of new phenomena. One can discover physical things (such as previously undiscovered elements, species, or stars), ideas (such as new concepts that lead to new theories), or understanding (such as new and improved understanding the behavior of natural phenomena). Discovery is different from invention and innovation. *See also* Innovation, Invention.
Discovery-Push Innovation	*See* Linear Innovation Models.
Discovery Research	*See* Research and Development.
Disembodied Technology	*See* Acquisition of Technology.
Disruptive Innovation	An innovation that displaces an earlier commercial product, process, or service and significantly changes (disrupts) a market niche, or even an entire market. Disruptive innovations are unexpected in the marketplace and are different from evolutionary or incremental innovations. Sometimes termed *Catalytic Innovation, Disruptive Technology, Discontinuous Innovation, Fundamental Innovation, Radical Innovation, Technology-Push Innovation, Transformational Innovation*, or *Revolutionary Innovation.* Disruptive innovation is usually very R&D intensive and may require discovery research. Examples: The electronic calculator made the slide rule obsolete; the electric typewriter made the manual typewriter obsolete. *See*

	Figures 1, 2–4. *See also* Innovation, Design-Driven Innovation, Evolutionary Innovation, Incremental Innovation, Radical Technological Transition.
Disruptive Technology	*See* Disruptive Innovation.
Distribution	In business, distribution refers to the process of making commercial products, processes, or services available to customers, either directly or through an intermediary. *See also* Marketing, Business Development, Sales.
Distribution Channels	A summary of the pathways available to get a product to the customers and/or end-users. *See also* Market Analysis. Reference [6].
Divergent Thinking	Part of Creative Problem-Solving. *See* Creative Thinking Models.
"Does it pencil?"	This phrase refers to the question of whether a new product or service concept has a sound business basis, meaning whether its cost and selling price will yield enough profit for it to be feasible. The phrase "*Back of the Envelop Calculation*" is sometimes used with the same meaning (although this phrase can also refer to purely technical, approximate calculations). Reference [6].
Domestic Product of Industry	(DPI) A measure of the economic production of industry within a region, given by the gross domestic product (GDP) less the production value of non-industrial goods and services, such as those provided by governments, not-for-profit organizations, and academia. *See also* Gross Domestic Product.
Downstream Innovation	*See* Upstream Innovation.
DPI	*See* Domestic Product of Industry.
Drucker, Peter (Ferdinand) (1909–2005)	An Austrian-born American management consultant and author. He is associated with the development of modern (post-1950) management principles. Among his influential writings are those on innovation, such as his 1985 book *Innovation and Entrepreneurship*.

Dual Ladder — *See* Technical Ladder.

Dual-Use Technologies — Technologies developed for application in one sector that can be used or adapted for use in another sector. Examples include the use of defense or aerospace technologies in manufacturing or other sectors. In some usage, *Dual-Use Technologies* specifically refer to defense technologies that can also be used for nonmilitary applications.

Due Diligence — The process of systematically researching and evaluating the accuracy of a statement or the nature and status of a person, group, or organization. Examples include conducting due diligence on a financial statement, the financial status of an organization, the background of a potential new employee. The goal of due diligence is usually to ensure that the important information related to a potential decision is known before making that decision, in order to reduce the risk of making a poor decision. Where patents are concerned, due diligence sometimes refers to a legal principle that patent owners need to develop a product, process, or service out of the patented intellectual property in question, rather than simply use the patent as a barrier to prevent others from doing so.

Dynamically Continuous Innovation — *See* Evolutionary Innovation.

E

Early Adopter — A category of technology adopter in Rogers' diffusion of innovation model. Early adopters have sometimes been referred to as "*Consumer-Innovators*," although the term refers to their role as early adopters and not as innovators. Also sometimes referred to as "*Visionaries*" or "*Creators*." *See* Technology Adoption Lifecycle. *See* Figure 6. *See also* Rogers, Everett M.

Early Growth Phase — *See* Venture Capital.

Early Majority — A category of technology adopter in Rogers' diffusion of innovation model. Also sometimes referred to as "*Pragmatists*," "*Critics*," or "*Collectors*." *See* Figure 6. *See* Technology Adoption Lifecycle.

Early Modern Innovation — *See* Innovation (Early Modern).

Early Stage Company — *See* Start-Up Capital.

Early Stage Technology — A technology concept that is beyond the speculative concept level and is either being developed into something feasible or has been demonstrated to be feasible. At this stage of development, the technology does not have to be practical, efficient, or cost effective. In terms of Technology Readiness Levels, a demonstrated proof of concept is beyond TRL 2 but may not yet fully satisfy TRL 3. *See also* Proof of Concept, Speculative Concept, Technology Readiness Level.

Earnings Before Interest, Taxes, Depreciation, and Amortization — (EBITDA) A financial measure intended to enable comparisons of companies' operating profitability by eliminating the effects of interest payments, tax jurisdictions, asset depreciation, and takeover activities. Some related terms include earnings before interest and taxes (EBIT), earnings before interest, taxes, and amortization (EBITA), earnings before interest, taxes, and depreciation (EBITD), and earnings before interest, taxes, depreciation, amortization, and restructuring or rent costs (EBITDAR).

DOI 10.1515/9783110429176-006

Ease of Entrepreneurship Index	A blended indicator calculated based on such factors as barriers to competition, regulatory and administrative opacity, and administrative burdens. Example: The Conference Board of Canada Ease of Entrepreneurship Index. *See* Innovation Indicators and Table 4.
EBIT	Earnings before interest and taxes. *See* Earnings Before Interest, Taxes, Depreciation, and Amortization.
EBITA	Earnings before interest, taxes, and amortization. *See* Earnings Before Interest, Taxes, Depreciation, and Amortization.
EBITD	Earnings before interest, taxes, and depreciation. *See* Earnings Before Interest, Taxes, Depreciation, and Amortization.
EBITDA	*See* Earnings Before Interest, Taxes, Depreciation, and Amortization.
EBITDAR	Earnings before interest, taxes, depreciation, amortization, and restructuring or rent costs. *See* Earnings Before Interest, Taxes, Depreciation, and Amortization.
Eco-Industries	As defined by the OECD and Eurostat, eco-industries produce products, processes, or services that measure, prevent, limit, minimize, or correct environmental damage to water, air, and soil, or otherwise deal with problems related to ecosystems. *See also* Eco-Innovation.
Eco-Innovation	*See* Ecological Innovation.
Ecological Innovation	The ideation, development, and deployment of improvements in environmental protection and/or contribute to sustainable development. Also termed *Eco-Innovation*. Example: A new recycling program. This is usually a form of non-commercial innovation, but technological innovation in this area occurs as well. Reference [33]. *See* Innovation (Non-Commercial). *See also* Eco-Industries.
Economic Cycle	The concept that, for various reasons, economies can cycle through periods of building, then rapid growth, then a plateauing and/or crisis, then a recession, and finally some kind of recovery or rebuilding. Economic Cycle is sometimes referred to as

	Business Cycle, or *Boom-Bust Cycle*. Several kinds of economic cycles have been identified, each with its own drivers, measures, and time-frames (periods).
Economic Innovation	The ideation, development, and deployment of improvements in economic models and processes. Example: Microcredit. *See also* Innovation. Reference [33].
Economy in Transition	*See* Competitiveness Drivers.
Ecosystem Innovation	Technological innovation that is identified, developed, and commercialized through a multiorganization partnership. This is similar to an R&D Alliance except that ecosystem innovation is aimed at developing and commercializing large technological innovation opportunities in a short time period, whereas R&D alliances are more focused on the R&D components and over moderate to long periods of time. Reference [59]. *Also termed* Multiparty Innovation.
Ecosystem Leader	*See* Business Ecosystem.
Educational Innovation	The ideation, development, and deployment of improvements in education models and processes. Example: E-learning. This is a form of non-commercial innovation. *See* Innovation (Non-Commercial). Reference [33].
Efficiency-Driven Economy	*See* Competitiveness Drivers.
Efficiency Enhancers	*See* Global Competitiveness Index.
Electronification of Innovation	A reference to the increasing use of electronic tools, such as expert systems, simulation modelling, and even artificial intelligence in 5th generation innovation practices. Reference [102]. *See also* Generations of Innovation, Non-Linear Innovation Models.
Embedded Knowledge	Knowledge that was once held only by a small number of specialists and/or experts but which has become embedded more generally in an organization's products, processes, services, or practices. *See also* Acquisition of Technology.

Embodied Technology	*See* Acquisition of Technology.
Emergent Novelty	*See* Innovation (20th–21st century).
Endless Frontier Research	*See* Third Mission.
Endless Transition Research	*See* Third Mission.
Endogenous Growth Theory	An economic growth theory that assumes economic growth is the result of internal (endogenous) rather than external (exogenous) factors. There are various versions of this theory, but the main internal growth factors are investments in human capital, technological innovation, and knowledge as they relate to productivity. As a result, the theory also gives importance to such things as industrial business strategies and government policies insofar as they affect the three principal drivers. Sometimes referred to as the "New Growth Theory." The Endogenous Growth Theory is different from external (exogenous) economic growth theories, such as the Solow-Swan Growth Model, which are based on the accumulation of physical capital and growth of the labor force, although both kinds of theories assume a strong connection between continuing technological innovation and overall economic growth. References [60, 61, 62]. *See also* Solow-Swan Growth Model, Exogenous Variable.
Endogenous Variable	*See* Exogenous Variable.
End User	The ultimate user of a product, process, or service. Such a user may or may not have any significant expertise in the nature of the product, process, or service, and the term is sometimes meant to imply a specific lack of such expertise. The purchaser of a product, process, or service is not necessarily an end user, which is an important consideration in the development of marketing strategies.
Engineering Prototype	A working model of a product or process that has been built carefully enough for use in testing and demonstration of important design parameters. An engineering prototype may not be

built to full scale, but will be designed to be scaled-up. *See also* Field Test, Prototype. Reference [6].

Engineering Research — *See* Scientific and Engineering Research.

Entrepreneur — A person that starts and manages a business, assuming all or most of the financial risks involved. *See* Reference [5]. *See also* Intrapreneur.

Entrepreneurial University — A university that has incorporated a "third mission," beyond those of teaching and research, which is to engage with other organizations in the development of practical uses of new knowledge ("*Mode 2 Research*"). Such a university can play a larger role in an innovation ecosystem. The entrepreneurial university is frequently associated with research and/or technology parks, incubators, and/or start-up and spin-off companies. In some definitions (including that of OECD) the entrepreneurial university is also itself associated with an entrepreneurial culture and can play a role in helping to develop entrepreneurs. The evolution of the entrepreneurial university with its third mission has also been referred to as the result of the "*second academic revolution*," the *first academic revolution* being the evolution of the research university, with its second mission, from the teaching university. References [63, 64]. *See also* Modes of Science, Innovation Ecosystem, Triple-Helix Model, Quad-Helix Model.

Envelope S-Curve — *See* S-Curve.

Equity — A term referring to the fractional ownership of a business corporation. It is sometimes used to refer to a specific percent of ownership in a business, sometimes to a number of shares (or the value of a number of shares), and sometimes to the total number or value of shares of a business.

Equity Capital — *See* Capital.

Equity Financing — *See* Financing.

Era of Ferment — A reference to the early stages in a product/process/service lifecycle, in which a new product is introduced to the marketplace, with associated business development, design improvements, and early to rapid sales growth as the product finds a successful market niche. The related terminology for the latter phases of the life-cycle is the "*Era of Incremental Change,*" in which sales plateau and then decline, as the market niche becomes saturated, and competition is mostly based on price. Reference [65].

Era of Incremental Change — *See* Era of Ferment.

Etzkowitz, Henry (1940 – Present) — An American sociologist known to the innovation world for his work on university-industry interactions and for his co-development (with Leydesdorff) of the Triple-Helix Model curve, a sociological model that was developed to describe the roles and intersections of governments, universities, and industry in advancing knowledge-based economies and which by extension has been applied to the advancement of economies based on innovation. He was also the originator of the "Entrepreneurial University" concept and co-founder of the Triple Helix International Conference Series (since 1996). *See* Triple-Helix Model. *See* Figure 8. *See also* references [63, 87, 150, 152, 153].

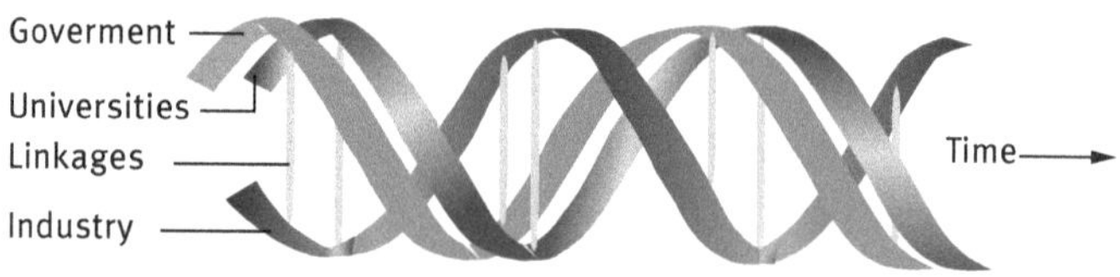

Figure 8: Illustration of the "Triple-Helix" Model of a Regional Innovation System. The vertical bars are drawn to illustrate the existence of continuing linkages along the development pathway.

Evolutionary Innovation — An innovation that competes with an earlier commercial product, process, or service but does not significantly change (disrupt) a market, or even a market niche. Evolutionary innovations usually significantly help companies enhance their competitive positions in the marketplace. Also termed *Dynamically Continuous Innovation*, or *Sustaining Innovation*. Sometimes termed *Continuous Innovation* although this term is probably better suited to *Incremental Innovation*. The terms *Architectural Innovation* and *Modular Innovation* are probably best considered to be forms of Evolutionary Innovation. Evolutionary innovation is usually very applied R&D intensive. *See* Figures 1 and 4.

See Architectural Innovation; Modular Innovation. *See also* Innovation, Disruptive Innovation, Incremental Innovation; Abernathy-Clark Model; Henderson-Clark Model.

Ex Ante — A Latin term meaning "from before." In business, the term *ex ante* usually refers to forecasts. For example, an *ex ante* estimate would be a forecast rather than a post-event, or "actual," estimate of financial or production results. In commercial Innovation, the term is sometimes used to refer to things, activities, or processes that are dealt with before an event. For example, several collaborators and/or their activities could be synchronized *ex ante* (i.e., before) an event of some kind. The opposite of *ex ante* is *ex post*, meaning "from after."

Exclusive License — *See* License Agreement.

Exit — *See* Exit Strategy.

Exit Strategy — A plan for when and how to leave a position. The position could be an ownership position, such as in the ownership of a start-up company, and the timing could be triggered by the achievement of pre-set objectives or the occurrence of unexpected risks or losses beyond pre-set tolerances. The plan could be as simple as selling shares, or might be more complex, and it could involve a phased transition.

Exogenous Variable — A variable whose origin is external to the system or model under consideration and which is independent of the system or model. Example: In classical economics (before the Schumpeter era), technological progress was generally assumed to be exogenous, meaning that it was assumed to be solely determined by noneconomic forces. The opposite of exogenous is endogenous.

Experimental Development — *See* Research and Development.

Explicit Knowledge — *See* Knowledge.

Explorers — (Innovation Strategy) A characterization of one of four kinds of organizational approaches to innovation strategy. "*Explorers*" tend to be in emerging, rapidly expanding markets, or redefining their markets, but in either case to be in markets demanding speed and agility. They also tend to be focused on customers, partners, and markets generally for new insights and opportunities, users of rapid prototyping and testing of large numbers of relatively small ideas, and innovating through a diffuse, customer-facing process. Jaruzelski and Dehoff have referred to such organizations as "*Need Seekers*" [19]. *See* Innovation Strategy Mapping.

(Technology Adoption) A category of potential customer or technology adopter in a psychological model of technology adoption. *See also* Technology Acceptance, Technology Acceptance Model, Technology Readiness Index, Technology Adoption Lifecycle.

Ex Post — See *Ex Ante*.

Ex Situ — In science and engineering, the Latin term *ex situ* generally refers to an aspect of a reaction or process taking place away from where it normally occurs or was created. *See also In Situ*.

External Patent Applications — (Innovation Indicator) An indicator of the number of practical inventions with international potential that are being protected is the number of external patent applications per 10,000 personnel in the labor force of an economy. *See* Innovation Indicators and Tables 4 and 8.

F

Factor-Driven Economy	*See* Competitiveness Drivers.
Fail Better	*See* Fail Fast.
Fail Cheap	*See* Fail Fast.
Fail Early	*See* Fail Fast.
Fail Fast	A common business mantra, especially with regard to product development, entrepreneurs, and start-up companies. Two key concepts associated with this phrase are: (1) not to fear failure but learn from it, and (2) that it is better to learn quickly if a new product, process, or service is not going to be successful so that it can either be improved or abandoned in favor of a new one. A goal could be to experience one or more small failures from which useful learnings can be obtained, increasing the probability of achieving a significant success, rather than risking a single large failure that might not be survivable. Also termed "Quick to Fail." Other nuances that build upon the same philosophy include: "Fail Cheap," "Fail Small," "Fail Often," "Fail Early," "Fail Better," "Fail Well," "Fail Forward." *See also* 'Works Like' Model.
Fail Forward	*See* Fail Fast.
Fail Often	*See* Fail Fast.
Fail Small	*See* Fail Fast.
Fail Well	*See* Fail Fast.
False Negative Error	*See* Decision-Making Errors.
False Positive Error	*See* Decision-Making Errors.

DOI 10.1515/9783110429176-007

False Summit — The phenomenon in a research or development process when a problem is solved and it is then discovered that there is another previously unrecognized problem yet to be solved before the real goal(s) of the process can be achieved. This is an analogy to a false summit in mountaineering, in which on ascent it appears that the summit of a mountain is in sight, but once reached, it turns out that it was simply a ridge or another feature of the mountain that had prevented sight of the true summit. The "*false summit effect*" refers to the psychological impact of reaching and recognizing a false summit, which is usually some level of discouragement. In early-stage innovation processes, the false summit effect occurs quite often because, by definition, not enough is known about the technological solution or how to reach it until the development is fairly well advanced.

False Summit Effect — *See* False Summit.

Fast Follower — A person or organization that takes advantage of a new idea, technology, invention, or technological innovation (developed by a "first mover") and implements a copy or an adaptation of it. A "*Fast Follower*" strategy (also termed a "*Second-to-Market Strategy*") has a goal of being an early participant in the marketplace as a new product, process, or service evolves "up" its life-cycle trajectory. Example: In the videotape industry, Sony was a first mover with its Betamax technology, while RCA and Matsushita were fast followers with their VHS technology. *See also* Adopt and Adapt, First Mover, Late to Market, Market Segmentation, Reverse Engineering.

Fast Innovator — *See* Time to Market.

Father of TRIZ — *See* Altshuller.

FBE — *See* Fuzzy Back-End.

Fear of Innovation — Some public fears related to technological innovation are that it may create and set loose upon society monsters (the "*Frankenstein Hypothesis*"); job losses and unemployment due to any or all of mechanization, process efficiency improvements, and robotics ("*Technological Unemployment*"); technologies that are advanced and marketable but environmentally and/or socially unsustainable; the driving or determining by technology of a society's social structure and cultural values ("*Technological Determinism*"). Reference [66].

Feasibility Study — An evaluation of a proposed initiative or business concept that is aimed at identifying strengths, weaknesses, opportunities, and threats (SWOT), the potential for success, and the resources needed to further develop the initiative or business concept. Feasibility studies are often used to help decide whether to proceed with an initiative or business concept. *See also* SWOT.

FEF — Front-End Fuzziness. *See* Fuzzy Front-End.

FFE — *See* Fuzzy Front-End.

FFF Capital — Friends, Family, and Fools Capital. *See* Seed Capital.

Field Demonstration — *See* Demonstration.

Field Pilot — A test of an engineering prototype of a new or improved product, process, or service under conditions of actual field, industrial plant, or market operation. This is usually the next step in commercialization beyond laboratory, proof-of-concept, or working model testing. Field pilots may be conducted at full commercial scale or at a reduced scale. Sometimes termed *Field Test, Pilot Plant, Pilot Test, Plant Pilot, Plant Test*, or simply *"Pilot." See* Figure 7. *See also* Clinical Trial, Demonstration, Engineering Prototype, Prototype, Scale-Up.

Field Test — *See* Field Pilot.

Fifth-Generation Innovation Model — The Systems Integration and Networking Model. *See* Generations of Innovation, Non-Linear Innovation Models.

Fifth-Generation of Technology Foresight — *See* Generations of Technology Foresight.

Fifth Wave — *See* 5th Wave.

Financing	The acquisition of capital. Some common forms in innovation are *Bootstrapping*, in which funds are generated internally from other income, *Debt Financing*, in which funds are borrowed, or *Equity Financing*, in which funds are obtained through the sale of a share in ownership.
Firm-First	*See* World-First.
First Academic Revolution	*See* Second Mission, Entrepreneurial University.
First-Generation Innovation Model	The Technology-Push Model. *See* Generations of Innovation, Linear Innovation Models.
First-Generation Nanotechnology	*See* Generations of Nanotechnology.
First-Generation of Technology Foresight	*See* Generations of Technology Foresight.
First Industrial Revolution	*See* Technological Ages.
First Mission	The first mission of a university is teaching. The second mission is research. The third mission of an "*Entrepreneurial University*" involves participation by universities in working with governments and industry to enable economic growth (via innovation) and social progress. *See also* Entrepreneurial University, Modes of Science, Innovation Ecosystem, Triple-Helix Model, Quad-Helix Model.
First Mover	A person or organization that is first to market a new product, process, or service, or the first to establish itself in a given marketplace. Also termed "First to Market," "Originator." A "*Market Leader*" strategy (also termed a "*First-to-Market Strategy*") has a goal of being the first to introduce new products, processes, or services into the marketplace. Example: In the videotape industry, Sony was a first mover with its Betamax technology, while RCA and Matsushita were fast followers with their VHS technology. *See also* Fast Follower, Late to Market, Market Segmentation.

First to Market — *See* First Mover.

First Wave — *See* 1st Wave.

Fixed Assets — *See* Assets.

Foresight — A systematic process involving environmental and horizon (futures) scanning aimed at anticipating future events in order to be able to develop strategies to encourage, prevent, change, or simply manage the impacts of such events. Sometimes termed "Futures Research". The principal kinds of future events anticipated by foresight activities usually involve technological, economic, environmental, political, social, and/or ethical (TEEPSE) futures. Innovation foresight (compared with technology foresight) has to look further than the development of new technologies and into their commercialization (this is also termed "*Innovation System Foresight*"). Two terms that occur frequently in foresight processes and scenarios are "*Wild Cards*" and "*Weak Signals.*" In foresight scenarios, Wild Cards are events that are perceived to be of low probability of occurrence but have potentially high impact should they actually occur. Also termed "*Outliers,*" "*Black Swan Events,*" or simply "*Black Swans.*" In foresight scenarios, Weak Signals are unclear observables that provide a warning of the probability of possible future events, including Wild Card events. Reference [67].

Formal Knowledge — *See* Knowledge.

Formative Stage Company — *See* Start-Up Capital.

Forward Innovation — *See* Reverse Innovation.

Foster, Richard N. — An American engineer and business consultant known to the innovation world for his adaptation of the sigmoid S-curve to applied research and technology development and to product life-cycles. This is described in his 1986 book, *Innovation: The Attacker's Advantage*. The S-curve is sometimes termed a Foster Curve. *See* S-Curve. Reference [26].

Foster's Curve — *See* S-Curve (Technology).

Fourth-Generation Innovation Model	The Integrated Model, *See* Generations of Innovation, Non-Linear Innovation Models.
Fourth-Generation Nanotechnology	Molecular nanotechnology. *See* Generations of Nanotechnology, Molecular Nanotechnology.
Fourth-Generation of Technology Foresight	*See* Generations of Technology Foresight.
Fourth Industrial Revolution	*See* Technological Ages.
Fourth-Pillar Organization	An organization that works to enable and/or assist with the innovativeness and competitiveness of companies, by working with industry, government, and academia (which represent the other three "pillars"). Such organizations use these linkages to help companies develop and deploy new commercial product, processes, and/or services. 4th Pillar Organizations are usually either government-owned corporations or not-for-profit corporations. Examples include research and technology organizations (RTOs), industry associations, and economic development organizations. 4th Pillar Organizations are also examples of "*Intermediary Organizations.*" Reference [68]. *See also* Intermediary Organization, Knowledge Intensive Business Services, Innovation Ecosystem.
Fourth Wave	*See* 4th Wave.
Frankenstein Hypothesis	*See* Fear of Innovation.
Frascati Manual	An Organisation for Economic Co-operation and Development (OECD) document providing recommended methods for collecting and interpreting data on research and development (R&D). The counterpart to the Frascati Manual is the Oslo Manual, which is concerned with data on innovation. Reference [7]. *See also* Oslo Manual.
Friends, Family, and Fools Capital	(FFF Capital) *See* Seed Capital.
Front-End Fuzziness	(FEF) *See* Fuzzy Front-End.

Frugal Engineering — *See* Frugal Innovation.

Frugal Innovation — This term generally refers to reducing the cost of an existing product, process, or service. Depending on the usage, there may be an implication that certain features have been removed for a particular market niche, that performance has been reduced, and/or that quality has been reduced. Also termed "*Inclusive Innovation.*" The terms "*Jugaad Innovation*" (using meagre resources) or "*Gandhian Innovation*" (affordable and sustainable) are sometimes used to describe Frugal Innovation with respect to the Indian market. Part of the process involved is sometimes termed "*Frugal Engineering.*"

FTE — *See* Full-Time Equivalent.

Full-Time Equivalent — (FTE) A unit of employment that recognizes fractional employment in order to give a better representation of workforce capacity than, for example, only recognizing full-time employment. As a result, the numbers of FTEs in an organization, or assigned to a program, may be very different from the number of people so employed or assigned. Example: the number of FTE researchers in an organization or assigned to a program.

Fundamental Innovation — *See* Disruptive Innovation.

Fundamental Research — *See* Research and Development.

Future Pull — Any kind of creative thinking, planning, or product/process/service development that is influenced mostly by a sense or vision of future benefits, needs, or demands could be said to be influenced by "*future pull.*" This term is sometimes used to provide contrast with the terms "*technology push*" and "*market pull.*"

Futures Research — *See* Foresight.

Fuzzy Back-End — (FBE) The final phase of new product/process/service development, when there is a very high degree of uncertainty about marketing and sales, and how best to manage these processes to achieve a commercially successful conclusion to the entire process. The term was coined to contrast with the "*Fuzzy Front-End*" phase. *See also* Fuzzy Front-End.

Fuzzy Front-End (FFE) The early phase of new product/process/service development, when there is a very high degree of uncertainty about the nature of the future product, whether it will be technically successful, what it will cost in terms of time and resources, and whether it will be likely to be commercially successful. The fuzzy front-end phase is typically from product conception through to the point where there is a validated product concept, before a decision is made to proceed with the product development phase. Also termed "*Upfront Phase*." The uncertainties inherent in this phase are sometimes referred-to as "*Front-End Fuzziness*," or FEF. *See also* Prototype, Time to Market, Fuzzy Back-End.

G

GAAP — *See* Generally Accepted Accounting Principles.

Gamma Release — *See Alpha* Test.

Gamma Test — *See Alpha* Test.

Gandhian Innovation — *See* Frugal Innovation.

Gartner Hype Cycle — *See* Technology Hype-Cycle.

Gated Process — *See* Idea-to-Launch Process.

GBAORD — Government Budget Appropriations or Outlays for R&D. *See* Government R&D Expenditures.

GCI — (1) *See* Global Competitiveness Index.
(2) Global Connectedness Index. *See* Connectedness Index.

GDP — *See* Gross Domestic Product.

GDP *per capita* — Gross Domestic Product per person, usually referring to a country or region within a country. *See also* Competitiveness Drivers.

Generally Accepted Accounting Principles — (GAAP) A set of standardized procedures and rules for recording and reporting corporate financial information.

Generations of Innovation — Rothwell identified five generations of models of technological innovation, also termed the "Rothwell Models of Innovation. Reference [102]. The first two are linear models of innovation, while the other three are non-linear models:
- 1st Generation: Technology-Push Model,
- 2nd Generation: Market-Pull Model,
- 3rd Generation: Coupling Model,

DOI 10.1515/9783110429176-008

- 4th Generation: Integrated Model,
- 5th Generation: Systems Integration and Networking Model.

See also Linear Innovation Models, Non-Linear Innovation Models, Generations of Technology Foresight.

Generations of Nanotechnology

The development and evolution of nanotechnology has been divided into four stages, or generations. First-generation nanotechnology refers to passive nanostructures created to carry out a single, specific task. This includes many formulated products having nanoscale components and products involving surface coatings. Second-generation nanotechnology refers to active nanostructures created to carry out more than one task, sequentially and/or in parallel. This includes nano-scale actuators, sensors, and drug delivery devices. Third-generation nanotechnology refers to the expected emergence of functioning nanosystems comprised of many (i.e., thousands of) interacting components. Fourth-generation nanotechnology refers to the expected emergence of nanosystems designed and constructed at a molecular level (molecular nanosystems). Such systems are predicted to work with hierarchical systems within systems, like living human or animal cells, and/or like nano-scale robots. *See also* Molecular Nanotechnology.

Generations of Technology Foresight

A set of descriptions of major advances in the evolution of technology foresight methodologies since the beginning of the 1990s. They are based in part on Rothwell's [102] five generations of models of technological innovation. First-generation foresight focuses on the prediction and diffusion of technologies, and often involves the "technology-push" model of innovation. Second-generation foresight adds to this the marketplace and the activities of the participants in technology development, be they industry, intermediary, government, and/or academic. This level of sophistication often involves the "market-pull" model of innovation. Third-generation foresight adds to this interactions with society, and often involves a "coupling" model of innovation (such as the "chain-linked" model). Fourth-generation foresight adds to this the distributed roles of the involved parties in innovation systems and often involves the "integrated" model of innovation. Fifth-generation foresight adds to this structural and policy issues and often involves the "*system integration and networking model*" model of innovation. References [69, 70]. *See also* Generations of Innovation.

GERD — *See* Gross Domestic Expenditure on R&D.

GIL — *See* Government Innovation Lab.

Ginarte–Park Index — An indicator of the relative strength of (level of protection provided by) a country's intellectual property laws. The index is based on the sum of five components of a country's patent laws, each of which is given a score of between 0 and 5. The components are breadth of coverage, participation in international patent agreements, protection against loss, degree of patent enforcement, and length of patent term. References [71, 72]. *See also* Patent, Patent Protection.

Global Competitiveness Index — (GCI) An indicator of a county's economic competitiveness that is calculated and regularly published by the World Economic Forum. The principal components of the GCI are "basic requirements," covering institutions, infrastructure, and the economic environment; "efficiency enhancers," covering higher education and training, the labor market, technological readiness, and market size; and "innovation and sophistication factors," covering business sophistication and capacity to innovate. Reference [36]. *See also* Competitiveness Drivers.

Globalized Regional Innovation System — *See* Regional Innovation System Types, Table 6.

Global MNE — Global Multinational Enterprise. *See* Multinational Enterprise.

Globalization — *See* Internationalization (Product, Process, or Service).

Global Multinational Enterprise — (Global MNE) *See* Multinational Enterprise.

GNERD — *See* Gross National Expenditure on R&D.

GNP — *See* Gross National Product.

Gompertz Curve	A technology S-curve model. The Gompertz logistic curve model relates the technology maturity (*y*) to time (*t*) as: $y = y_{\infty} \cdot e^{\alpha \cdot \exp(-\beta t)}$, where the maximum value for *y* is given by y_{∞} and the shape of the S-curve is determined by the adjustable parameters α and β. In this model, the growth curve is not symmetrical about the inflection point. Instead, it sharply increases up to the inflection point, after which the rate of growth slows. *See also* S-Curve, Pearl Curve.
Go-to-Market Strategy	A strategy by which an organization connects with, and delivers products, processes, and/or services to, its target customers.
GOVERD	*See* Government Intramural Expenditure on R&D.
Government Budget Appropriations or Outlays for R&D	(GBAORD) *See* Government R&D Expenditures.
Government-Financed GBAORD	*See* Government R&D Expenditures.
Government-Financed Gross Domestic Expenditure on R&D	(Government-Financed GERD) *See* Government R&D Expenditures.
Government Innovation Laboratory	(GIL) Not a laboratory but a physical or virtual space in which government officials, sometimes with other stakeholders such as business or community people, meet to conceptualize, discuss, and propose innovative government practices, policies, and/or services. For an example of a virtual GIL, see Reference [73]. Innovation in this context is not technological innovation, but is a form of organizational and/or social innovation. *See also* Innovation (Non-Commercial).
Government Intramural Expenditure on R&D	(GOVERD) Research and development expenditure made within the government sector. *See also* Government R&D Expenditures.

Government R&D Expenditures There are several ways to measure or estimate government spending on R&D. Two commonly used measures are *Government-Financed Gross Domestic Expenditure on R&D* (Government-Financed GERD) and *Government Budget Appropriations or Outlays for R&D* (GBAORD). Government-financed GERD is simply the portion of GERD that is financed by government. GBAORD is obtained from government budgets by identifying all the budget items involving R&D and estimating their R&D content in terms of funding. GBAORD therefore represents government R&D spending intentions rather than actuals. *See also* Government Intramural Expenditure on R&D, Gross Domestic Expenditure on R&D, Gross National Expenditure on R&D, Frascati Manual. Reference [7].

Government R&D Intensity (Innovation Indicator) An indicator of the R&D investments made by governments is the ratio of government budget appropriations or outlays for R&D to gross domestic product (GBAORD/GDP). *See* Innovation Indicators and Table 4.

Grant Capital *See* Capital.

Grassroots Regional Innovation System *See* Regional Innovation System Types, Table 6.

Great Surges *See* Kondratieff Waves.

Grey Space *See* White Space Mapping.

Gross Domestic Expenditure on R&D (GERD) The total expenditure on R&D performed within a country or region during a given period. GERD includes R&D performed within the country or region and funded from abroad, but it does not include payments for R&D conducted outside the country or region. GERD is calculated by adding the expenditures in each of the four performing sectors: business, not-for-profit, government, and academic. *See also* Gross National Expenditure on R&D, Frascati Manual, Government R&D Expenditures. Reference [7].

Gross Domestic Product (GDP) The total economic productivity of a region, such as a country, or of an industrial sector during a specified period of time, usually a year or quarter of a year. GDP is different from business sales in that it takes into account the sales value of the goods and services produced less the value of the goods and services consumed during the production. Thus, GDP can potentially increase with increased production, increased sale prices, and/or reduced costs of production. The difference in GDP per capita between a country or region and some specified peer or peers is termed the "*Prosperity Gap*." *See also* Domestic Product of Industry, Macroeconomics. *See also* Gross National Product.

Gross National Expenditure on R&D — (GNERD) The total expenditure on R&D financed by a country's institutions during a given period, including the financing of R&D performed outside of the country. GNERD is calculated by adding the expenditures in each of the four performing sectors: business, not-for-profit, government, and academic. *See also* Gross Domestic Expenditure on R&D, Frascati Manual, Government R&D Expenditures. Reference [7].

Gross National Product — (GNP) The total economic productivity of the people and companies of a country, regardless of where the actual production took place, during a specified period of time, usually a year. GNP may be lower or higher than GDP depending on how much of a country's production takes place outside of its borders and how much of other countries' production takes place within its borders. *See also* Gross Domestic Product.

Gross Rate of Replication — *See* Technological Capacity.

Growth Curve — *See* S-Curve.

Growth Model — *See* Solow-Swan Growth Model.

H

Henderson-Clark Model — An innovation model that defines forms of evolutionary innovation and distinguishes between an organization's product component knowledge and its architectural knowledge (i.e., knowledge of the linkages among product components). In the *Henderson-Clark Model,* organizations having low degrees of both component and architectural knowledge are most likely to achieve modest incremental innovations at best, whereas those with high degrees of both component and architectural knowledge are most likely to achieve disruptive innovations. In between these extremes fall two categories of organizations. One category represents organizations having a low degree of component knowledge but a high degree of architectural knowledge, and therefore are most likely to achieve *Architectural Innovations*, in which improvements are made to the linkage(s) between the components in a product, but not to the components themselves. The other category represents organizations having a low degree of with architectural knowledge and a high degree of component knowledge, and therefore are most likely to achieve *Modular Innovations*, in which improvements are made to one of more of the components in a product, but not among the linkages between those components. The Henderson-Clark Model has been used to explain how incumbent organizations in a marketplace can achieve substantial innovations that fall short of being disruptive. Reference [20]. *See* Figure 3. *See also* Incremental Innovation; Evolutionary Innovation; Disruptive Innovation; Abernathy-Clark Model; Platform Innovation; Radical Technological Transition.

HERD — Higher Education Expenditure on R&D

Heroic Theory of Discovery (or Invention) — *See* Multiple Discovery Theory.

Hesitators — A category of potential customer or technology adopter in a psychological model of technology adoption. *See also* Technology

DOI 10.1515/9783110429176-009

Acceptance, Technology Acceptance Model, Technology Readiness Index, Technology Adoption Lifecycle.

Heterodox Paradigm
The Heterodox Paradigm was coined by Michael Storper in 1997 [74] to represent a new way of thinking about economic development (and innovation), in terms of regions, organizations, and technologies and their linkages and interdependencies (as embraced in the *"Holy Trinity" Model*). Storper's Heterodox Paradigm is an example of *heterodox economics* coupled with technology development and sociology. Heterodox economics refers to approaches to economics that go beyond, or at least fall outside of classical or conventional economics, such as socio-economics or eco-economics. *See also* "Holy Trinity" Model.

High-Tech
See High-Technology.

High-Technology
This usually refers to the technological sophistication of a new product, process, or service, although the term has also been applied to entire industries, sectors, and even regions. Technological sophistication, in this context, is often assessed based on direct (and sometimes also indirect) R&D intensity. Four categories of technological sophistication are often distinguished, based on some measure of R&D intensity, such as R&D investments divided by value-added and/or gross production values:
- High-Technology (also termed High-Tech), such as aerospace, pharmaceuticals, and instruments,
- Medium-High-Technology (also termed Medium-High-Tech), such as chemicals, electrical equipment, and motor vehicles,
- Medium-Low-Technology (also termed Medium-Low-Tech), such as refined petroleum products, metals and metal products, and shipbuilding, and
- Low-Technology (also termed Low-Tech), such as food, pulp and paper, and textiles.

Such categorizations are both generalized and relative. A "*High-Tech*" industry can produce "*Low-Tech*" products, and *vice versa*. "*High-Technology*" is sometimes used simply as a synonym for "*Advanced Technology*." Reference [75].

High Technology Complex
See Innovation Ecosystem.

Holding Period Return *See* Return on Investment.

Holland, Maurice (1891–1981) An American engineer and research manager. He was the first head of Engineering and Industrial Research at the US National Research Council and seems to have published the first description of the steps involved in the "*research cycle*" leading to innovation (in 1928). His work was the forerunner of the first linear model of innovation, advanced by Maclaurin in the 1940s. He was also the founder of the US Industrial Research Institute (in 1938). References [76, 77]. *See also* Maclaurin.

"Holy Trinity" Model A regional innovation system model advanced by Michael Storper in 1997 to describe the roles and intersections of territories, technologies, and organizations in linking regions, technologies, and production capabilities into innovation systems that generate new products and economic development [78]. Storper termed this way of thinking about economic development (and innovation) the Heterodox Paradigm. The first-order interactions in this model would produce economic technological systems, regional technology systems, and regional productions systems, all of which combined would lead to a stream of new products entering the marketplace. *See also* Heterodox Paradigm, Innovation Ecosystem.

Horizontal Thinking A synonym for Brainstorming. *See* Creative Thinking Models.

Hype-Cycle *See* Technology Hype-Cycle.

Hyper-Innovation (Pace of Innovation) The simplest definition of hyper-innovation is innovation conducted at a rapid pace. Depending on the specific usage, there is frequently an associated connotation of rapid innovation that is driven, or at least enabled, by multiple interconnections among, people, organizations, information technologies, and markets.

(Amount of Innovation) In some contexts, the term hyper-innovation is used to mean "too much" innovation. Example: An organization that has been highly successful with one or more major technological innovations could become vulnerable to a culture shift toward hubris, undermining the kind of culture that originally enabled the innovation(s). *See also* Hypo-Innovation.

Hypo-Innovation — This term generally refers to "too little" innovation. In this context, the prefix *hypo* means "less than normal." Example: An organization that has not produced significant technological innovations and is being outpaced by its competitors for this reason could be considered to suffer from hypo-innovation. *See also* Hyper-Innovation.

I

I18N — "I-eighteen letters-N" referring to a specific kind of internationalization. *See* Internationalization (Product, Process, or Service).

ICT — *See* Information and Communication Technology.

ICT Investment Intensity — The ratio of ICT sector expenditures to gross domestic product in an economy. *See* Innovation Indicators and Table 4.

Idea Generation — *See* Ideation.

Idea Sun Bursts — *See* Spider Diagrams, *under* Creative Thinking Models.

Ideation — The process of idea generation whether through divergent thinking, brainstorming, Camelot Scenarios, or any other approach. Organized and/or facilitated ideation is sometimes used to originate ideas for new theories, applications, solutions, processes, or products. *See also* Breakthrough, Innovation Project.

Idea-to-Launch Process — Any of a number of processes that have been developed to guide the managed evolution of an idea into a commercial enterprise, product, process, or service. Also termed "*Gated Process*." Examples of idea-to-launch processes for new product development in an organization include the Stage-Gate® process and the Product and Cycle-Time Excellence (PACE) Process. *See* Stage-Gate® Product Development Process, Technology Stage-Gate Process, and Figure 9.

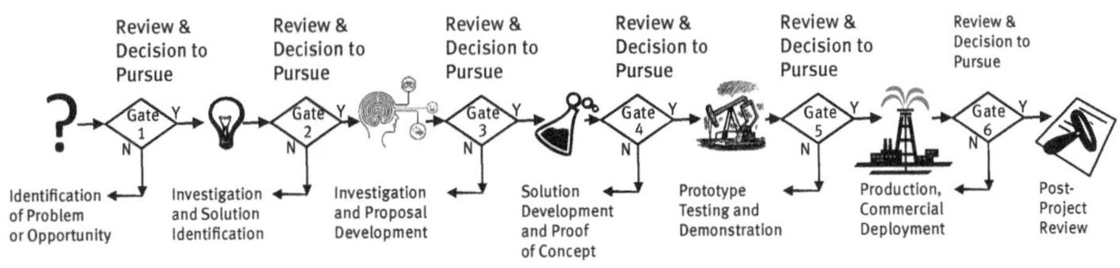

Figure 9: Illustration of a Stage-Gate® Process for the Development of a New Petroleum Industry Process.

DOI 10.1515/9783110429176-010

ILO	*See* Industry Liaison Office.
Imitative Innovation	*See Shanzhai* Innovation.
Imitator	*See* Innovator.
Impact Multiplier	A number that quantifies the indirect effects of an economic impact. The principle is that an economic impact such as in increase in product sales will be multiplied in an economy through increased spending on the jobs, goods, and services needed to produce those products. Multipliers tend to be estimates because they attempt to account for indirect effects (sometimes termed *spillover effects*) that may not be directly or easily measurable. Multipliers are expressed as a ratio between a measure of a total effect and a measure of the direct effect that caused it.
Impacts	(Innovation) The ultimate consequences of the outputs from an organization or its employees, especially in the context of the mandate and/or mission of the organization. Examples: Sales and adoption of a new product, process, or service (output) might result in significantly increased customer satisfaction, revenues, net income, jobs, exports, or economic activity. *See also* Activities, Inputs, Outcomes, Outputs, Reach, Metrics.
Implementation	*See* Technology Deployment.
Inactive	*See* Laggards.
Inbound Innovation	*See* Inbound Open Innovation.
Inbound Open Innovation	An aspect of the open innovation process in which the flow of technology or other information is into an organization from the outside. Example: licensing-in of technologies. Also termed Inbound Innovation. *See also* Open Innovation, Outbound Open Innovation.
Inclusive Innovation	*See* Frugal Innovation.

Incremental Innovation — An innovation that represents a small improvement, or a series of small improvements to an existing product, process, or service. Incremental innovations are often more or less continuous and may be required to maintain a competitive position in the marketplace. Sometimes termed *Continuous Innovation, Core Innovation*, or *Market-Pull Innovation.* Sometimes termed *Sustaining Innovation,* but this term is probably better suited to *Evolutionary Innovation.* Incremental innovation is usually very problem solving- and adaptation-intensive. *See* Figures 1, 3, and 4. *See also* Innovation, Disruptive Innovation, Evolutionary Innovation, Technology Mudslide Hypothesis.

Incrementalist Strategy — *See* Rationalist Strategy.

Incremental-Radical Dichotomy — The dilemma faced by large, successful, incumbent companies in an established marketplace that tend to find it extremely difficult to implement disruptive innovations because, by their very nature, they require the creation of new technological knowledge and/or resources that render the old ones obsolete, and they involve the introduction of such radically new products, process, or services into the marketplace that they render the previous ones obsolete – including those of the innovating company. The dilemma is whether to risk disrupting a well-established and successful business, or line of business, by introducing a new product that may or may not be successful. Also referred to as the *Innovator's Dilemma. See also* Abernathy-Clark Model, Innovation Paradoxes (Innovation System Paradoxes).

Incubation — *See* Business Incubation.

Incubator — *See* Business Incubator.

Indigenous Innovation — (Self-Determined) Innovation processes of any kind conducted for and/or by a group of people in a region or country for themselves. Also termed "*Zizhu chuangxin,*" meaning self-determined innovation. Example: National innovation programs.

(Indigenous People) Innovation, often social innovation, processes conducted for and/or by Indigenous people. A related meaning is social innovation processes that are informed and potentially influenced by the application of indigenous (traditional) knowledge as guided by elders, and hence the wisdom of ancestors.

(Global Transfer) A process of making use of technologies transferred from other, more advanced economies, to develop new or improved technologies "at home." Example: China's 2006 indigenous innovation strategy. *See also Shanzhai* Innovation.

Indirect Indicators — *See* Innovation Indicators.

Individual Contributor Ladder — *See* Technical Ladder.

Induced Innovation — An older term that used to be used with reference to drivers of innovation such as knowledge-induced (i.e., technology-push) innovation and demand-induced (i.e., market-pull) innovation. Such terminology was commonly used in the 1960s. Reference [79]. *See also* Linear Innovation Models.

Industrial Design — A grant of exclusive rights on the visual features of an original, manufactured item by a government. Such features generally include some combination of features of shape, pattern, or configuration, but not the method or materials used in the manufacturing. *See* Invention, Intellectual Property.

Industrial Innovation — *See* Innovation (20th–21st century). *See also* Maclaurin.

Industry Liaison Office — (ILO) A technology transfer and/or industry liaison office at an academic institution such as a university or polytechnic. The principal focus of such offices is usually to facilitate the transfer of the knowledge gained through discovery research to those that would undertake applied research and development, build links between academic researchers and industry, assist faculty members with patenting, and negotiate licensing agreements, particularly the latter. At universities, these are sometimes termed "*University-Industry Liaison Offices*" (UILOs).

Information Age or Society — *See* Technological Ages.

Information and Communication Technology (ICT) A rapidly growing area that includes communication devices, applications, and services. Examples range from satellite sensing and communication systems, to network hardware and software, to devices such as radios, televisions, phones, and computers. ICTs formed the basis for key developments in the 5th Kondratieff Wave of disruptive innovations. *See also* 5th Wave, Kondratieff Waves.

Information-Based Economy *See* Knowledge-Based Economy.

Information Economy *See* Knowledge-Based Economy.

Information Hierarchy *See* Wisdom Hierarchy.

Information Society *See* Knowledge-Based Economy.

Infrastructural Investment Cycle *See* Kuznets Cycle.

Infrastructure of the Information Society *See* Internet of Things.

In-Kind A phrase meaning "with something similar." In business this phrase usually refers to a non-monetary payment or investment such as with goods or services, in which case the in-kind contribution is usually assigned an equivalent monetary value.

Innovation (Early modern) In the 1500s through 1800s, innovation meant "*introducing novel change,*" particularly with regard to religious and/or political change. In these early times, the connotation of the term innovation was negative: as it was frequently meant to imply that the changes were unwanted, unnatural (apart from the natural order of things), revolutionary, and/or dangerous. In contrast, the terms "*reformation*" or "*restoration*" were frequently used to describe positive, moderate, natural-order-restoring changes. Reference [1].

(20th–21st century) From the 1930s to the present day, a technological, or commercial, meaning of the term has been widely used, in which innovation involves introducing novelty to the commercial marketplace through new products and services, and in which innovation has a positive, but still disruptive, connotation. This meaning was defined by Schumpeter in the 1930s as the conversion of ideas and knowledge into commercially successful products and services. Here, "successful" means meeting the needs of customers, in a way that encourages them to take up the new approach and have it diffuse through the marketplace. Whereas an invention is the first occurrence of an idea or concept for a new product or process, innovation involves taking it into commercial practise. In this sense, some innovations can be described as successfully commercialized discoveries or inventions. The terms "*technological innovation*[3]," "*commercial innovation,*" or "*industrial innovation*" are often used to distinguish the 20th–21st century, or Schumpeterian definition of innovation from the Early Modern definition. An early synonym, attributed to A.P. Usher [80] is "*emergent novelty*." References [3,4]. *See also* Discovery, Invention, Top-Down Innovation, Upstream Innovation.

Several types of technological innovation have been defined based on the nature of the technology, hence *Product Innovation, Process Innovation,* and *Service Innovation.* Examples of process innovations are Henry Ford's automobile assembly line manufacturing process, and FedEx's overnight long-distance parcel delivery process. *Marketing Innovation* and *Organizational Innovation* refer to improvements in the practice of marketing or the operations of an organization, respectively, in ways that ultimately translate into improved sales and/or margins for the products, processes, or services that the organization is selling in the marketplace. Reference [8]. Several types of innovation have been defined based on outcomes. *See* Disruptive Innovation, Evolutionary Innovation, Incremental Innovation. *See also* Adjacent Innovation.

---

3 The term "*technological innovation*" came into use in the 1950s, representing a merger of the work of Maclaurin and Schumpeter and may have been coined by Maclaurin, who frequently referred to "technological change."

(Non-Commercial) In 21st century usage, the earlier meaning of the term innovation as introducing novel change has again become widely used, but this time with a positive connotation, and it has been extended to realms beyond those of religion, politics, and technology, such as organizational processes, marketing, and social structures. An example is the ideation, development, and deployment of new and improved internal processes within an organization, even if these have little or no connection to commercialization or the marketplace. The terms "*non-commercial innovation*" and "*non-technological innovation*" are sometimes used to distinguish such non-Schumpeterian innovation. Thus, the OECD *Oslo Manual* definition of innovation would be taken to include both technological and non-commercial innovation. *See also* Organizational Innovation, Government Innovation Lab.

Innovation 2.0 — A variation on the "*system integration and networking model*" (SIN Innovation Model) that involves parallel development integrating horizontal linkages among communities, governments, industry, research organizations, and universities, with the aim of creating and deploying disruptive innovations that drive economic growth, improved quality of life, and reduced resource use and environmental impacts [81]. Example: a community's adoption of a smart traffic flow-adjusting system that responds to real-time traffic and air quality sensor readings [81]. *See also* Non-Linear Innovation Models.

Innovation Accelerator — *See* Business Accelerator.

Innovation Activities — *See* Activities. *See also* Inputs, Outcomes, Outputs, Metrics, Reach, Impacts.

Innovation and Sophistication Factors — *See* Global Competitiveness Index.

Innovation at the Bottom of the Pyramid — (BOP Innovation) Technological innovation (usually products) developed in and targeting emerging economies. Also termed "*Resource-Constrained Innovation.*" *See also* Reverse Innovation.

Innovation Barrier Any challenge or obstacle to the successful process of innovation up to and including successful technology commercialization. Such barriers could be technological, financial, manufacturing, cultural, or regulatory, for example. Such barriers are more specific, however, than the general requirements to conduct R&D, experimental development, prototyping, pilot testing, demonstration, preproduction, financing, marketing, sales, and deployment. By definition, all innovation barriers have to be successfully identified and managed before innovation (commercialization) is possible. *See also* Breakthrough, Camelot Scenario.

Some common "traps" that can create barriers to innovation include the "*Performance Trap*" and the "*Commitment Trap*." The "*Performance Trap*" can occur when a company is not performing well and focuses on short-term tactics at the expense of long-term strategy (sometimes referred to as the "*Short-Term Mindset Trap*"), or when a company is performing well and focuses on its current successes at the expense of other opportunities (sometimes referred to as the "*Business Model Trap*"). The "*Commitment Trap*" can occur when a company fails to commit to a new opportunity and extensive research, analysis, and/or testing inhibit or prevent progress, or when a company over-commits to an opportunity that no longer appears prospective, inhibiting, or preventing retreat. Reference [82].

Innovation Black Box *See* Black Box.

Innovation Bridge A reference to the function of technological innovation in providing a bridge between the world of knowledge, new discoveries, and ideas on the one hand, and the marketplace of commercialized products, processes, and services, on the other hand.

Innovation Cluster *See* Cluster (Innovation Cluster).

Innovation Continuum (Technology) A way of describing and/or mapping the progress of technology development illustrating the stages involved and the degrees of technology maturity that they represent. *See* S-Curve (Technology) and Figure 10. *See also* Technology Readiness.

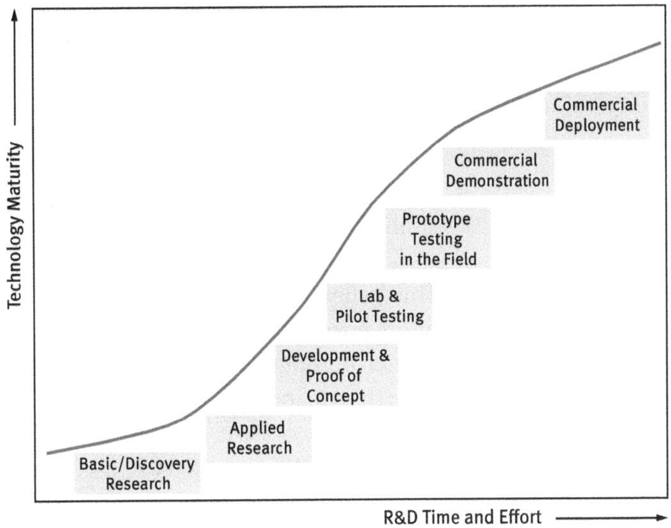

Figure 10: Illustration of a Technology S-Curve.

(Degree of Innovation) A way of describing and/or mapping degrees of innovation and their impact on an economy. *See* S-Curve (Innovation) and Figure 11. Also termed Innovation S-Curve.

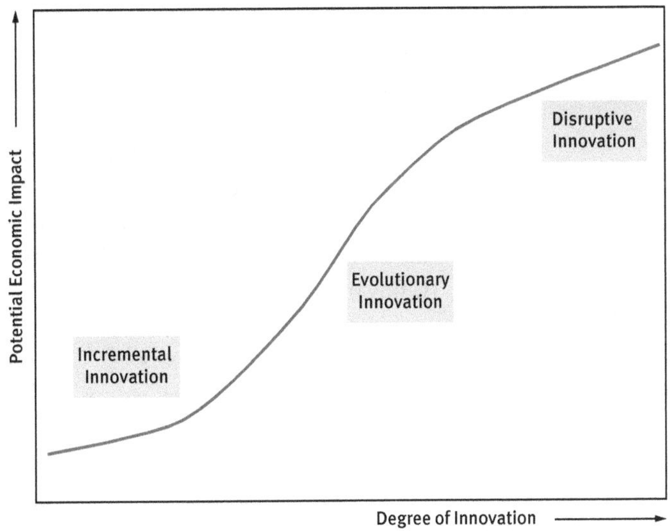

Figure 11: Illustration of an Innovation S-Curve.

Innovation-Decision Process The process by which a potential customer evolves from first becoming aware of a technological innovation, to forming an opinion on it, to reaching a decision on whether or not to purchase it, to acting on the decision, to confirming the purchase decision. Rogers has summarized these as: (1) knowledge, (2) persuasion, (3) decision, (4) implementation, and (5) confirmation [31].

Innovation Deficit This generally refers to either a rate comparison or a return-on-investment comparison. In the former case, it is when the rate of introduction of new technological innovations is less than that of another specified organization, region, or country. In the latter case, it is when the rate of introduction of new technological innovations is less than what would be taken to be an acceptable rate of return given the investments made in support of innovation. In either case, this term is usually used subjectively. Where it is used quantitatively, it is usually not innovation that was measured but some other measure, such as economic activity or patents granted, with an implicit assumption that the measure is directly linked to innovation.

Innovation Diffusion Model *See* Rogers.

Innovation Dissonance The tensions that can arise among people and teams when an organization is trying to encourage and promote technological innovation and encounters any of the several so-called *innovation paradoxes* by which the drivers of innovation come up against elements of excess, competition, organizational momentum, or even strategy. Depending on the nature and degree such tensions could be considered to be productive or unproductive. Reference [83]. *See also* Innovation Paradoxes (Innovation Process Paradoxes).

Innovation-Driven Economy *See* Competitiveness Drivers.

Innovation Ecosystem (General) An innovation ecosystem includes all of a region's public and private sector people and organizations whose activities include any or all of developing, enabling, producing, or diffusing innovations. These people or organizations constitute innovation ecosystem entities because of their capabilities and regardless of whether or not they actually interact with each other. Examples of innovation ecosystem entities include individual inventors and entrepreneurs, for-profit companies, not-for-profit companies, government organizations, and academic organizations. An *Innovation System* is a network of innovation ecosystem people and/or entities that interact with each other to develop, enable, and/or produce innovations. *See* Figure 12. *See also* Business Ecosystem, Innovation System Theory, Innovation Ecosystem Models, Innovative Regional Cluster, Triple-Helix Model, Quad-Helix Model, Quad Model.

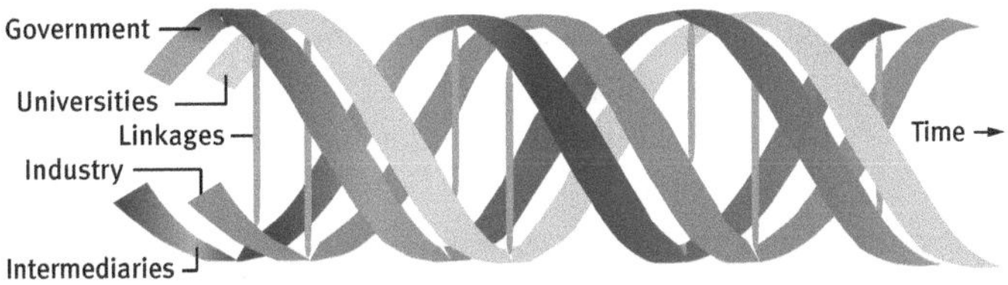

Figure 12: Illustration of the Quad-Helix Model of a Regional Innovation System. The vertical bars are drawn to illustrate the existence of continuing linkages along the development pathway.

(National and Regional) The term *National Innovation System* (or National System of Innovation, NSI) refers to a nation's innovation system, and *Regional Innovation System* to a region's. A national or regional innovation system comprises the network of public- and private sector innovation-related institutions and their mutual interactions. Reference [84]. Regional Innovation Systems have also been referred to as: *Innovative Milieux*, *High Technology Complexes*, *Technopolis Complexes*, *Technopoles.* Clusters can be considered a special kind of regional innovation system. *See also* Regional Innovation System Types.

Innovation Ecosystem Entities — *See* Innovation Ecosystem.

Innovation Ecosystem Models — Any of several models of innovation ecosystems that focus on the interacting roles of multiple institutions in the process of creating technological innovations. The kinds of institutions considered vary with the model but generally include three or four of governments, universities, intermediary organizations, and industry. These models generally attempt to describe how discovery, knowledge-generation, invention, product development, market launch, on so on depend on interactions among institutional players, involving elements of control and/or incentives, knowledge generation, and wealth generation. Examples include *"Holy Trinity" Model*, *Triple-Helix Model*, *Quad-Model*, *Quad-Helix Model*, *N-Tuple Helix Model*, among others. *See* Figures 8 and 12. *See also* Holy Trinity Model, Innovation Ecosystem, Quad-Helix Model, N-Tuple Helix Model, Triple-Helix Model.

Innovation Expenditures — The expenditures associated with innovation, including *Research and Development*. *See also* Government R&D Expenditures, Gross Domestic Expenditure on R&D, Gross National Expenditure on R&D, Activities, Research and Development.

Innovation Foresight — *See* Foresight.

Innovation Funnel — *See* Product Development Funnel.

Innovation Gap — The gap between desired and actual innovation performance as measured or perceived by stakeholders of some kind. The innovation performance gap could apply to regions, countries, sectors, or companies, for example. The stakeholders could be governments, organizations, companies, leaders, employees, or constituents, for example. Also termed *Commercialization Gap* or *Technology Gap*.

Innovation Gap Theory — A theory that a country, region, industry, or company that achieves new innovations (products, processes, or services successfully introduced into the marketplace) will have a competitive advantage over their competitors until the gap is closed. In the case of countries or regions the theory is sometimes extended by assuming that other countries or regions will have to import the new innovations until they can be replicated or substituted within their own regions. In this latter sense, an innovation gap can stimulate international trade. Also termed Commercialization Gap Theory or Technology Gap Theory.

Innovation Impacts — *See* Impacts. *See also* Activities, Inputs, Outcomes, Outputs, Metrics, Reach.

Innovation Incubator	*See* Business Incubator.
Innovation Indicators	Regions, countries, and organizations, in their efforts to measure and benchmark innovation performance, tend to use a combination of direct and indirect (proxy) indicators. Some of the indicators are "*lagging indicators*," meaning that they represent the past (such as measures of new products, process, and services already introduced into a marketplace). Other indicators are "*leading indicators*," meaning that they are used to represent possible future innovation performance (such as measures of resources and activities directed toward technological innovation). References [85, 86]. *See* Tables 4, 6, 8.
Innovation in Innovation	Reference coined by Etzkowitz to the broadening of the term innovation to include improvements in the way organizations interact in order to create the conditions for innovation and/or to enable innovation to occur [87]. *See also* Entrepreneurial University, Innovation Ecosystem Models, Triple-Helix Model, Quad-Helix Model.
Innovation Inputs	*See* Inputs. *See also* Activities, Outcomes, Outputs, Metrics, Reach, Impacts.
Innovation Intermediary	*See* Intermediary Organization.
Innovation Intermediation	*See* Intermediary Organization.
Innovation Killer	Anything that contributes a barrier to creating innovation. Cynthia Rabe describes the ways in which having too much deep knowledge, or expertise, in an organization or team can be an innovation killer [88].
Innovation Management	An organizational function that coordinates efforts to accomplish technological innovation. Such efforts could include planning, organizing, staffing, and/or leading processes involving, for example, market research, product research and development, manufacturing, and marketing. Innovation management is different from Technology Management. *See also* Intellectual Property Portfolio Mining, Technology Management.

Innovation Metrics *See* Metrics. *See also* Activities, Inputs, Outcomes, Outputs, Reach, Impacts.

Innovation Models *See* Linear Innovation Models, Non-Linear Innovation Models.

Innovation Outcomes *See* Outcomes. *See also* Activities, Inputs, Outputs, Metrics, Reach, Impacts.

Innovation Outputs *See* Outputs. *See also* Activities, Inputs, Outcomes, Metrics, Reach, Impacts.

Innovation Paradoxes The literature contains quite a few concepts of paradox inherent to innovation. These generally refer to either the pros and cons of pursuing technological innovation in the face of organizations or products/processes/services that are already successful ("*Innovation System Paradoxes*"), or else the challenges that may be inherent within the process of innovation itself ("*Innovation Process Paradoxes*"). As such they are usually more about risks than paradoxes in a literal sense, but the literature references are most often to "paradoxes."

(*Innovation System Paradoxes*) One system paradox could be called the "*Success/Failure Paradox*," in which an organization's single-minded focus on "success" (and lack of tolerance for "failure") can blind it to the fact that it actually needs to be able to tolerate, and even embrace failures in order to achieve success through technological innovation. Termed the "*innovation paradox*" by Farson and Keyes [89], who argue that it could be limiting or counter-productive to even think in terms of successes and failures. Another system paradox could be called the "*Success Paradox*," in which an organization becomes so successful, through technological innovation, that it becomes blinded to the need for more or new innovations, causing it to miss out on breakthrough innovations that, to its ultimate detriment are left to other, more nimble companies. Example: Nokia famously killed its smart phone because they were so deeply invested in "dumb phones." Termed the "*innovation paradox*" by Davila and Epstein [90]. Another version of this is the paradox an organization faces when it recognizes that developing a breakthrough innovation could lead it to future success but that the pathway could involve destroying its currently successful products, processes, or services. This is termed the "innovator's dilemma" by Christensen [14].

(*Innovation Process Paradoxes*) Some examples of innovation process paradoxes are included here: In the "*Cultification Paradox*," the single-minded pursuit of innovation is taken to such extremes that it overrides focus on other key activities such as sales and customer focus (i.e., too many pilots, too many specification changes, too many new products, etc.) or that it causes changes to be made "for the sake of innovation" where there is no need for innovation at all (i.e., changing process that are already effective and efficient). In the "*Collaborative Compromise Paradox*," multiple groups are brought together to collaborate on a complex innovation challenge but the competing needs and priorities of such groups lead to compromises that reduce their combined effectiveness, making the whole less than the sum of its parts, rather than more. In the "*Invention Within Convention Paradox*," inventions that could lead to innovations are developed but then cannot be developed further because the organization's operational processes are not flexible enough to accommodate changes. In the "*Novelty Paradox*," novelty is needed in product/process/service innovation, but a novel new product or a novel change to an existing product may not be perceived as an improvement by customers. Even worse, the introduction of a new or changed product is often accompanied by discontinuation of the previous one which could lead to the loss of existing customers who are unable to purchase the old product and do not want the new one. In the "*Productivity Paradox*," investing in the development of a new product, process, or service with an expectation of increased productivity may lead to very positive results in other areas but either no increase, or even a loss in productivity itself. An example is the "*Solow Computer Paradox*:" that increased investment in information technology could cause labor productivity to decrease instead of increase[4]. In the "*Organizational Paradox*," if an attempted technological innovation fails or performs poorly in the marketplace, or if an attempted organizational innovation fails, the impacts on the organization, such as supply chain problems or reduced spending budgets leading to reduced quality control (or reduced

4 In the early 1970s, many organizations began investing in computer systems with the expectation that they would enable labor productivity improvements (or even eliminate labor completely in some areas). However, although computers and information technology enabled many companies to achieve competitive advantage and market share, the desired labor productivity increases were largely unrealized, and in some cases, decreases were experienced.

customer service), for example, could lead to the loss of existing customers. This could be considered to be "*negative innovation.*" Reference [33, 83]. *See also* Innovation Dissonance, Negative Innovation, Solow Computer Paradox.

Innovation Park — *See* Research and Technology Park.

Innovation Performance Mapping — There are several systems for comparing the relative innovation performance and capabilities of organizations. Figure 13 shows an example in which organizations are rated as having both high innovation capabilities and performance (Red Giants), both low innovation capabilities and performance (Black Holes), high innovation capabilities but low innovation performance (Comets), or low innovation capabilities but high innovation performance (Shooting Stars). The latter category includes organizations that have achieved innovation success(es) but which do not have the ability to sustain them. Reference [91].

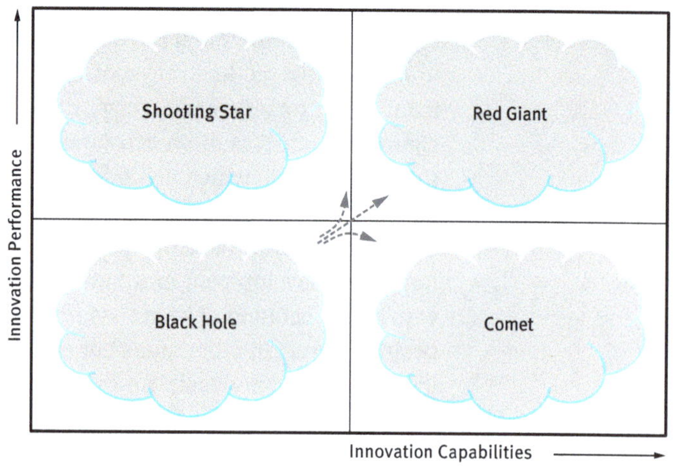

Figure 13: Illustration of Innovation Performance Mapping.

Innovation Pipeline — (I-Pipe) An idea pipeline, or gathering mechanism, intended to feed a new product development process. Generally attributed to Whirlpool. This is similar in concept to the Idea-to-Launch Process or Product Development Funnel (Process). *See also* Idea-to-Launch Process, Product Development Funnel, Stage-Gate® Product Development Process, Technology Stage-Gate Process. Reference [92, 93].

Innovation Process Paradoxes	*See* Innovation Paradoxes.
Innovation Project	A project whose goal is to develop a new idea, discovery, or invention into an innovation. This could be an individual or group initiative.
Innovation Reach	*See* Reach. *See also* Activities, Inputs, Outcomes, Outputs, Metrics, Impacts.
Innovation-Related Triple-Helix	*See* Triple-Helix Model.
Innovation Sandbox	*See* Sandbox.
Innovation S-Curve	*See* S-Curve (Innovation).
Innovation Snail	A term coined by Badulin for a technological product's development and business cycle, described as a linear series of 12 stages beginning with a product idea, and continuing through development, business planning, capitalization, production, market saturation, and then either declining sales or transitioning to another newer product. Badulin's sequence was drawn in an almost open-circular shape having the appearance of a snail's shell. Reference [112]. *See also* Product Life-Cycle Curve in the entry for S-Curve, Buying Hierarchy.
Innovation Strategy Mapping	An approach to categorizing organizations' innovation approaches, having four categories based on whether innovation is developed internally or externally, and how structured or informal the innovation process tends to be, *see* Figure 14. In this approach, the four categories are: "*Explorers,*" who tend to be in emerging, rapidly expanding markets, focused on customers, partners, and markets generally for new insights and opportunities, users of rapid prototyping and testing of large numbers of relatively small ideas, and innovating through a diffuse, customer-facing process. "*Architects,*" who tend to be in mature markets, with intensive capital and resource requirements and a centralized, structured approach to innovation, focused on customers and competitors for new insights

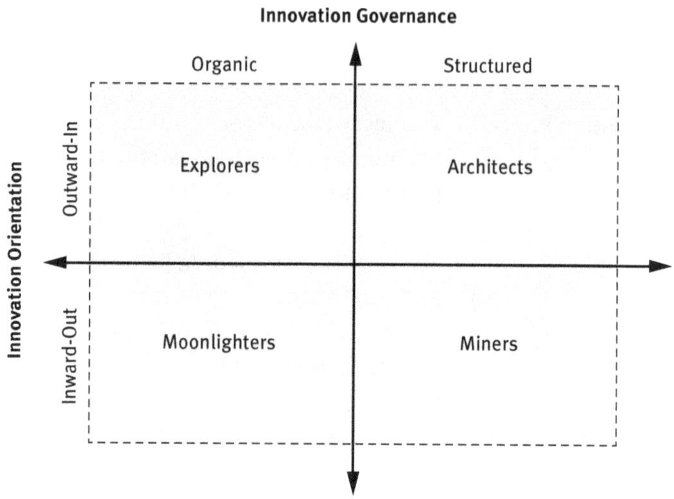

Figure 14: Illustration of Innovation Strategy Mapping (adapted from Swahney [94]).

and opportunities, outsourcers of development and prototyping, and innovating through a top-down, formal process. "*Moonlighters,*" who tend to be large companies with lots of technology and process expertise, that look inward to develop innovations based on internal ideas and know-how, focused on a strong innovation culture that supports internal experimental and development project initiatives, including internal "moonlighting," willing to fund internally generated innovation projects, and innovating through bottom-up processes. "*Miners,*" who tend to be very large companies with lots of technology and process expertise, very large internal business units, and strong silos, weak on internal innovation, necessitating the use of external innovation resources and/or organizations, focused on finding innovation opportunities within the organization, and innovating through formal, centralized processes working through the external innovation organization. Reference [94].

Innovation System — *See* Innovation Ecosystem.

Innovation System Entities — *See* Innovation Ecosystem.

Innovation System Foresight — (ISF) *See* Foresight.

Innovation System Paradoxes — *See* Innovation Paradoxes.

Innovation System Theory — The hypothesis that a nation's or region's innovation performance depends upon the breadth and depth of their innovation ecosystem and on the quality of the relationships and interactions among the innovation ecosystem entities. A corollary is that entities like government departments and agencies, research and technology organizations (RTOs), universities and colleges, and businesses, and even entire industries in a country or region need to interact effectively with each other in order to stimulate and/or enhance system innovation performance. *See also* Innovation Ecosystem.

Innovation Trap — *See* Innovation Barrier.

Innovation Union — An initiative launched by the European Union's (EU) economic growth strategy "Europe 2020." The aim of the EU's Innovation Union is to improve Europe's technological innovation performance, in order to achieve economic growth and jobs growth.

Innovation Voucher Program — Government programs that are intended to help small- and medium-sized enterprises (SMEs) become more productive and innovative by developing and launching new products, processes, or services into the marketplace, while building stronger linkages with research and development (R&D) providers (public and private). The vouchers are generally some kind of credit note for a certain amount of money that the SME can use to access expertise and technical solutions from R&D providers. Depending on the program the SMEs themselves may be required to contribute financially to the project as well. Examples of eligible services include applied research, development engineering, prototyping, product design and testing, intellectual property assessments and patenting, pilot and field testing, and market assments and advice. Such voucher programs have various names but they usually reference innovation, productivity, and/or competitiveness. Reference [95].

Innovative Firm — An organization that has introduced an innovation (to the marketplace) during a specified period of time.

Innovative Industrial Cluster	*See* Innovative Regional Cluster.
Innovative Regional Cluster	A geographically defined grouping of organizations of all sizes that comprise an industrial sector in which network relationships exist and are linked to associated academic, intermediary, and government organizations, and for which enhanced innovation performance have been attributed to the network as a whole. An "*Innovative Industrial Cluster*" is similar, but is not geographically defined or constrained. *See also* Innovation Cluster, Innovation Ecosystem.
Innovative Milieu	*See* Innovation Ecosystem.
Innovator	(Innovation) A person or organization that is the first to create an innovation, or at least is the originator of the critical conceptual elements of an innovation. Others making or adopting the same technological changes later are sometimes referred to as "*imitators.*" The term innovator is often incorrectly applied to discoverors or inventors. *See also* Discoveror, Inventor, Invention, Patent. (Technology Diffusion) A personal or organizational category of technology adopter in Rogers' diffusion of innovation model. Also sometimes referred to as "*Technologists*" or "*Creators.*" *See* Figure 6. *See* Technology Adoption Lifecycle.
Innovator's Dilemma	A term coined by Clayton Christensen, referring to his observation that the practices that allow a company to become a leader in a mainstream market tend to cause them to miss opportunities offered by disruptive innovations, and that pursuing such disruptive innovations requires practices that could undermine the company's current business position. The dilemma then becomes whether to protect the company's current business position or risk it in pursuit of a new potential business. Reference [14]. Also referred to as the *Incremental-Radical Dichotomy* in innovation. *See also* Abernathy-Clark Model, Innovation Paradoxes (Innovation System Paradoxes).
Inputs	(Innovation) Things that an organization or its employees receive or acquire in order to carry out a particular mission and/or achieve an objective. Example: Contract or grant funding, new

hires, purchases, or externally generated ideas or knowledge. *See also* Activities, Outcomes, Outputs, Reach, Metrics, Impacts.

In Silico	In science and engineering, the Latin-based term *in silico* generally refers to an experiment, calculation, or process simulation conducted by numerical (computer) simulation.
In Situ	In science and engineering, the Latin term *in situ* generally refers to an aspect of a reaction or process taking place where it normally occurs, as opposed to moving it to some other place. Examples include studying a phenomenon where it occurs, in a reaction mixture, or in a process vessel. *See also Ex Situ.*
Institute of Technology	*See* Polytechnic.
Institutional Innovation	*See* Organizational Innovation.
Intangible Assets	*See* Assets.
Intangible Knowledge	*See* Social Technology.
Integrated Circuit Topographies	A grant of exclusive rights on the three-dimensional features of an original integrated circuit design by a government. Such features generally include various layers of metals, insulators, and semi-conductors in an electronic circuit, but not the actual functions of the circuit. *See* Invention, Intellectual Property.
Integrated Innovation	A coordinated, systems-engineering approach to innovation in which scientific, engineering, technological, business, and sociological aspects are all considered throughout the complete innovation process. This would include all of what are sometimes termed "*Innovation Activities*" plus the "*Other Innovation Activities.*" *See also* Innovation, Research and Development, Activities, Other Innovation Activities.
Integrated Model of Innovation	*See* Non-Linear Innovation Models. *See also* Generations of Innovation.
Intellectual Capital	*See* Technological Capacity.

Intellectual Property	(IP) Almost any definable intellectual creation including products, processes, services, literary and artistic works, symbols, names, images, designs (including logos), and other forms of know-how. Some forms of IP are protected by holding them in secrecy, such as concepts, discoveries, and *trade secrets*, which can include non-patentable inventions. Some forms of IP are legally protectable under intellectual property law, such as by *contract, copyright, patent, trade dress*, or *trademark. See* Intellectual Property Rights. *See also* Invention, Technology.
Intellectual Property Mining	*See* Intellectual Property Portfolio Mining.
Intellectual Property Portfolio Mining	(IP Portfolio Mining) The scanning and evaluating of intellectual property (IP), including but not limited to patents, held within an organization and/or external to an organization. The purpose is to identify IP that is dormant, underutilized, or that could be utilized in new ways, as part of an organization's innovation management program. Classic examples are the identification of patented technologies from another country that are either not protected in the organization's home country or could be licensed for use in that country. Also termed *Intellectual Property (IP)Mining, Patent Database Mining, Patent Mining.*
Intellectual Property Rights	(IPR) Formal protection that is granted under a country's intellectual property law. Intellectual property rights that are protected by legislation include patents, trademarks, copyrights, industrial designs, integrated circuit topographies, and plant breeders' rights. Intellectual property can be held as a trade secret, but in this case there is no legislative protection. *See* Invention, Intellectual Property, Patent, Trademark, Copyright, Trade Secret, Industrial Design, Integrated Circuit Topographies, Plant Breeders' Rights.
Intelligent Materials	*See* Smart Materials.
Interaction Model	A term referring to, or a version of, the "*coupling model*" of the technological innovation process. *See* Non-Linear Innovation Models.
Interactive Model of Innovation	*See* Linear Innovation Models.

Interactive Regional Innovation System — *See* Regional Innovation System Types, Table 6.

Intermediary Organization — (General) In general, an intermediary organization functions between an organization and some or all of its stakeholders by providing some kind of service, such as a program function or technical assistance, for example. Brokers and delears are sometimes referred to as Technology Market Intermediaries (TMI). *See also* Knowledge-Intensive Business Services.

(Innovation) In the technological innovation world, an intermediary organization is one that functions between industry and the marketplace in a way that supports technological innovation processes. Also termed *Bridging Organization, Broker, Innovation Intermediary* (and sometimes *Superstructure Organization* or *Knowledge Processor*). Example: A Fourth-Pillar Organization such as a Research and Technology Organization (RTO). The specific functions provided by the intermediary organization could be almost anything that enables, assists, or leverages an innovation process, such as foresight, information/knowledge processing, research and development, testing, accrediting, and commercializing, for example. The process of conducting such activities is sometimes referred to as *Innovation Intermediation.* References [96, 97]. *See also* Fourth-Pillar Organization, Knowledge Intensive Business Services, Innovation Ecosystem Models, Research and Technology Organization.

International Corporation — *See* Multinational Enterprise.

Internationalization — (Corporation or Organization) The entry of an organization into operations that cross national borders. The term frequently refers to the entry of a company into sales of products, processes, or services in one or more international markets.

(Product, Process, or Service) Designing a product, process, or service so that it can be sold in multiple countries, or such that it can be readily adapted to be sold in multiple countries. This meaning of internationalization is also termed *I18N* (i.e., "I-eighteen letters-N"). The sub-process of actually adapting such a product, process, or service to a specific country, language, and/or culture is referred to as *Localization*. This meaning of the localization is also termed *L10N* (i.e., "L-ten letters-N"). It follows that an internationalized product, process, or service should be fairly easy to localize. The process of internationalizing a product, process, or service and then localizing it for different specific countries, continents, or cultures is sometimes referred to as *Globalization*.

Internationally Advanced Organization — An organization that stands out internationally by having globally unique, or at least globally uncommon, capabilities and/or products and services.

Internet of Things — (IoT) The networking of physical devices of all kinds ("*IoT Devices*") such that they can collect and exchange data, allowing remote sensing and controlling across a network. Example: smart thermostat systems that can be remotely monitored and controlled via Wi-Fi. The Internet of Things has been called "the infrastructure of the information society" and could enable broad system controls such as the "*smart grid*" (a smart electric power grid) and "*smart cities*" (using a broad information and communication technology system to manage a city's infrastructure and systems). *See also* Pervasive Computing.

Intrapreneur — A person that acts entrepreneurially within a larger company. For example, an intrapreneur might develop, launch, and lead a new line of business or a new business unit within an organization, drawing upon the capacity of the larger organization but acting with greater risk-tolerance and nimbleness than is usual for the organization. *See also* Entrepreneur.

In Vacuo — In science and engineering, the Latin term *in vacuo* generally refers to a process or procedure conducted in a vacuum.

Invasionary Technology — A technology that, upon commercialization, competes directly with one or more technologies that already dominate in the marketplace. Reference [6].

Invention — A new thing such as a new synthetic molecule, computer software program, device, or process. Some inventions are made public, some are described in materials that are copyrighted, some are held as trade secrets, and others are patented. A patentable invention is an invention that meets a patent system's requirements for novelty, usefulness, and significance. That is, to be patentable an invention must be original enough not to be obvious to someone skilled in the field of the particular invention, more than a minor improvement on the prior inventions in the field, and it must be useful (it is usually described as a solution to a problem) – which is usually demonstrated by a practical application (*Reduction to Practice*) of the concept. Most inventions are minor and/or "obvious" improvements ("*subinventions*") and not patentable. Most patented inventions do not have significant commercial value, so it typically requires considerable R&D and a multitude of inventions to realize a patentable invention that does have significant commercial value. Invention is different from discovery and innovation. *See also* Discovery, Innovation, Patent, Prior Art.

Invention Within Convention Paradox — *See* Innovation Paradoxes (Innovation Process Paradoxes).

Inventive Activity — Work that is aimed at creating inventions. Example: Much of Thomas Edison's work was specifically aimed at creating practical inventions that could be commercialized. Reference [98].

Inventive Potential — The total number of hypothetical inventions that could, in principle, be made by a member of an organization or society at a given point in time. Schmookler defined inventive potential as the number of inventions that a member of an organization or society could make at a given point in time with the talents that they have and the knowledge that anyone in the organization or society has available to them. Reference [98].

Inventor — A person that has created or co-created an invention. Where an invention is patented, there are rules for determining inventorship. *See also* Discoveror, Innovator, Invention, Patent.

Investment in Knowledge — The ratio of total knowledge investments, including higher education, R&D, and software, to gross domestic product in an economy. *See* Innovation Indicators and Table 4.

In Vitro	In science and engineering, the Latin term *in vitro* generally refers to a process or procedure conducted in an artificial environment such as a laboratory, as opposed to the "natural" setting. Example: growing biological cells in a laboratory culture, rather than within an organism (which is *in vivo*).
In Vivo	In science and engineering, the Latin term *in vivo* generally refers to a process or procedure conducted in its natural environment, as opposed to in an artificial environment. Example: growing biological cells in a living organism, rather than in a laboratory culture (which is *in vitro*).
IoT	*See* Internet of Things.
IoT Devices	*See* Internet of Things.
IP	*See* Intellectual Property.
I-Pipe	*See* Innovation Pipeline.
IP Mining	*See* Intellectual Property Portfolio Mining.
IP Portfolio Mining	*See* Intellectual Property Portfolio Mining.
IPR	*See* Intellectual Property Rights.
ISF	Innovation System Foresight. *See* Foresight.

J

J-Curve — In general, a J-curve is any graph or diagram in which plotted values first decrease for some time but ultimately reach a minimum and rise thereafter. In business, business lines, and in technological innovation, J-curves usually refer to revenues initially being characterized by growing annual costs (negative revenue) as a business struggles to gain sales and market share, or as research, development, and commercialization costs are incurred, respectively. In successful businesses, business lines, or technological innovations, there is ultimately reached a tipping point beyond which sales revenues offset costs and ultimately total revenues become positive and growing. Such curves stereotypically approximate the shape of the letter "J." J-Curves are also used in economics, with regard to trade balances for example. *See also* S-Curve.

Joiner — *See* Late Majority.

Jugaad Innovation — *See* Frugal Innovation.

Juglar Cycle — An economic cycle having a period of about 7 to 11 years, reflecting oscillations in the demands on and levels of investments in production facilities. Within this cycle, businesses experience first an expansion phase, then a crisis, followed by a recession, and ultimately a recovery phase. *See also* Economic Cycle, Kondratieff Waves.

Justified True Belief — *See* Knowledge.

DOI 10.1515/9783110429176-011

K

KBE — *See* Knowledge-Based Economy.

KETs — *See* Key Enabling Technologies.

Key Enabling Technologies — (KETs) A small set of technologies that are expected to be the most important "building blocks" for future technological innovation across all industrial sectors. These technologies are expected to play critical roles in the evolution and sustainability of leading-edge economies. Examples: Advanced Materials, Micro- and Nano-Electronics, Nanotechnology, Photonics, Industrial Biotechnology, and Advanced Manufacturing. Examples of industries for which KETs could lie at the heart of disruptive, or "game-changing," technological innovations include Aerospace, Agriculture, Automotive, Building Construction, Food, Healthcare, Mining and Minerals, Oil and Gas, Specialty Chemicals, and Textiles. Reference [99].

Keystone — *See* Business Ecosystem.

Key Technology — A mature technological innovation that has been widely accepted and adopted by customers, with essentially complete technology diffusion and market saturation. *See also* Pacemaker Technology; Incremental Innovation; Evolutionary Innovation; Disruptive Innovation.

KIBS — *See* Knowledge-Intensive Business Services.

KIS — Knowledge-Intensive Services. *See* Knowledge-Intensive Business Services.

Kitchin Cycle — An older economic cycle (circa 1920s) having a period of about 3 to 4 years, reflecting how decision making in companies was affected by the time required for information to flow. This factor has become less important as the nature and pace of information flows has dramatically improved. In modern usage, the Kitchin Cycle tends to refer more to the time to recognize that a market saturation has occurred, make decisions in response,

DOI 10.1515/9783110429176-012

adjust production and inventories, and/or wait for market demand to re-establish. *See also* Economic Cycle, Kondratieff Waves.

KM — *See* Knowledge Management.

Know-How — *See* Technology.

Knowledge — In the context of innovation knowledge refers to facts, information, and understanding that have been acquired through education and/or experience, with reasonable assurance of their validity. This has also been referred to as "*justified true belief.*" "*Tacit Knowledge*" refers to knowledge acquired through thinking and experience in a specific context, comprises cognitive (beliefs and viewpoints) and technical (skills and know-how) aspects, and is therefore difficult to transfer to others. "*Explicit Knowledge*" refers to knowledge that can be accurately described, codified, and recorded and is therefore easy to store and transfer to others. Also termed Codified or Formal Knowledge. "*Knowledge Codification*" refers to the conversion of tacit knowledge into explicit knowledge, permitting it to be accessed and used by others. *See also* Acquisition of Technology; Knowledge-Based Economy.

Knowledge Age or Society — *See* Technological Ages.

Knowledge-Based Economy — (KBE) An economic model referring to a society whose economy is largely based on a combination of knowledge workers, the acquisition, creation, dissemination, and utilization of knowledge (embodied knowledge, and codified knowledge), and the translation of such knowledge into learning, innovation, and economic development. In some usage, there is also an implication of broad access to and sharing of the knowledge (at least within the society under consideration). A knowledge-based economy is therefore more sophisticated than simply an "information-based" or "learning-based" economy (which only creates and disseminates information) or a "market-based" economy, and also different from a "political" economy. Also termed "Knowledge Economy" or, more broadly, Knowledge Society (in which economic health and growth can be translated into social health and growth). Similarly, an Information-Based Economy is also termed "Information Economy" or "Information Society." Reference [100]. *See also* Knowledge.

Knowledge Codification — *See* Knowledge.

Knowledge Economy — *See* Knowledge-Based Economy.

Knowledge Hierarchy — *See* Wisdom Hierarchy.

Knowledge-Induced Innovation — *See* Linear Innovation Models.

Knowledge Intensive Business Services — (KIBS) Almost any kind of business service that involves or is strongly reliant on sophisticated technological and/or other professional knowledge. The former, technological KIBS, is sometimes referred to as T-KIBS (e.g., science, engineering, R&D) while the latter, professional KIBS, is sometimes referred-to as P-KIBS (e.g., accounting, law, marketing). Sometimes simply referred to as Knowledge-Intensive Services (KIS). In some cases, KIBS organizations act as intermediary organizations. Brokers and dealers are sometimes referred to as *Technology Market Intermediaries* (TMI). *See also* Intermediary Organization.

Knowledge Intensive Services — (KIS) *See* Knowledge-Intensive Business Services.

Knowledge Management — (KM) The processes of any combination of identifying, developing, capturing, storing, using, sharing, and/or selling knowledge (embodied knowledge and/or codified knowledge). This can apply to an individual or to an organization. *See also* Knowledge-Based Economy.

Knowledge Processor — *See* Intermediary Organization.

Knowledge Society — *See* Knowledge-Based Economy.

Knowledge Transfer The process of capturing, organizing, and transferring knowledge from one part of an organization to another or from one organization to another. The term is often used to refer to kinds of knowledge that are historically and/or culturally embedded and difficult to articulate.

Knowledge Turn The amount of time required to develop an idea, concept, theory, or discovery into a commercialized innovation. Reference [9].

Kondratieff, Nikolai Dmitriyevich (1892–1938) A Russian economist and student of free markets who is best known for his theory that major capitalist economies tend to go through 50-year cycles in which an economic depression (of about 10 years), would be followed by a period of technical advances (of about 30 years), followed by a period of economic uncertainty (of about 10 years). In modern usage, these cycles are usually taken to have periods of about 40 to 60 years, with the periods declining in magnitude over time. Eponyms include Kondratieff Waves and K-Waves. In some works, his name is translated as Kondratiev. *See also* Kondratieff Waves.

Kondratieff Waves Waves or cycles through which major economies evolve. Nikolai Kondratieff proposed in 1925 that economics tend to go through 50-year cycles in which an economic depression (of about 10 years) would be followed by a period of technical advances (of about 30 years), followed by a period of economic uncertainty (of about 10 years) [101]. These cycles have at their heart periods in which a number of key technological developments (as well as social/political and economic) developments enable bursts of disruptive product or service innovations that fuel a sharp increase in industrial, and therefore economic growth. The waves of change also carry with them the seeds of the next wave so the cycles repeat, apparently with periods of approximately 40 to 60 years. Synonyms include Long Waves, Long Cycles, Supercycles, Great Surges, K-Waves, and Long Economic Cycles. Kondratieff is sometimes written "Kondratiev." References [11]. *See also* 1st through 7th Wave, Economic Cycle, Macroeconomics, Technological Ages, and Figure 15.

Kuznets Curve An older economic cycle (dating back to the 1930s) having a period of about 15 to 25 years, reflecting demographic changes, usually due to immigration or emigration waves, and their

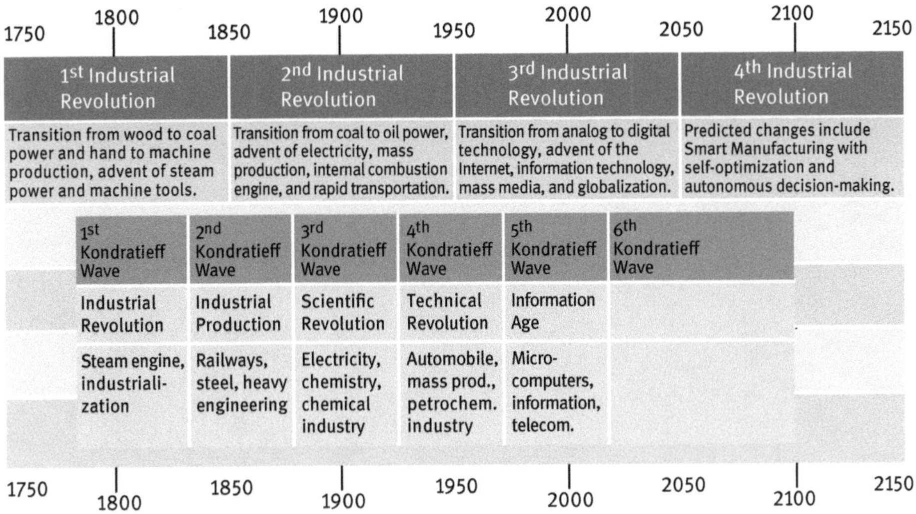

Figure 15: Illustration of Kondratieff Waves and the Industrial Revolutions.

effects on building construction. Also sometimes termed Kuznets Swing, Demographic Cycle, Building Cycle, Infrastructural Investment Cycles. *See also* Economic Cycle, Kondratieff Waves.

Kuznets Swing — *See* Kuznets Cycle.

K-Waves — *See* Kondratieff Waves.

L

L10N	"L-ten letters-N" referring to a specific kind of localization. *See* Internationalization (Product, Process, or Service).
Laggard	A category of technology adopter in Rogers' diffusion of innovation model. Also sometimes referred to as "*Late Mass,*" or "*Sceptics,*" or "*Inactives.*" *See* Figure 6. *See* Technology Adoption Lifecycle.
Lagging Indicators	*See* Innovation Indicators.
Late Majority	A category of technology adopter in Rogers' diffusion of innovation model. Also sometimes referred to as "*Conservatives,*" "*Joiners,*" or "*Spectators.*" *See* Figure 6. *See* Technology Adoption Lifecycle.
Later Stage Company	*See* Start-Up Capital.
Late to Market	In terms of innovation strategy, a person or organization that takes advantage of a new idea, technology, invention, or technological innovation (developed by a "first mover") and implements a specifically low-cost adaptation of it. A "*Cost Minimization*" strategy (also termed a "*Late-to-Market Strategy*") has a goal of producing a product, process, or service in the middle to late stages of a life-cycle trajectory. *See also* First Mover, Fast Follower, Market Segmentation.
Large-Size Enterprise	*See* Small- and Medium-Size Enterprise.
Late Mass	*See* Laggards.
Lateral Thinking	*See* Creative Thinking Models.
Law of the Minimum	*See* Liebig's Law of the Minimum.

DOI 10.1515/9783110429176-013

Leading-Edge Customer One of several kinds of stereotypical people that can play a critical role in the technological innovation process. Engaging prospective customers throughout the innovation process is often identified as a means of improving the success rate. An idealized leading-edge customer might be an early adopter of technology, an experienced user of previous innovative products, able to provide forward-looking advice on desired new products or product features or benefits, a potential tester of prototypes, and a source of potential post-launch improvement ideas. Reference [102]. *See also* Rainmaker, Product Champion, Technological Gatekeeper.

Leading Indicators *See* Innovation Indicators.

Lead Users Key customers that organizations consult for ideas that could be developed into new products, processes, or services. Sometimes termed Luminaries. This is somewhat similar to the concept of a Leading-Edge Customer.

Lean Innovation A technological innovation process that is systematic, focused on the things that add value, and eliminates waste. This is by analogy with the principles of "*Lean Manufacturing.*" Reference [103].

Learning Society *See* Technological Ages.

Learning-Based Economy *See* Knowledge-Based Economy.

LEPEST Analysis Legal, Environmental, Political, Economic, Social, and Technological Analysis. *See* Social, Technological, Economic, Environmental and Political Analysis.

Leydesdorff, (Louis André) Loet, (1948 – Present) A Dutch sociologist known for his work in the sociology of communication and innovation. He co-developed (with Eztkowitz) the Triple-Helix Model, a sociological model describing the roles and intersections of governments, universities, and industry in advancing knowledge-based economies, and which by extension has been applied to the advancement of economies based on innovation. *See* Triple-Helix Model. See also references [22, 113, 114, 154].

License	*See* License Agreement.
License Agreement	An agreement under which a person, group, or organization is entitled to make, use, or sell some form of intellectual property. Under an *Exclusive License* a licensee gains the sole right to such use, although there may be some kind of limitation such as geographic territory, technical field of use, product application, or time. Under a *Limited or Nonexclusive License,* a licensee gains a right – but not the sole right – to such use, although again there may be some kind of limitation. Payments made by a licensee for the right to use licensed intellectual property are called *royalties. See also* Licensing, Cross-License, Royalty-Free Licence.
Licensing	The legal process of contractually granting an intellectual property license. *See also* License Agreement, Cross-License, Royalty-Free License.
Liebig's Law of the Minimum	Named for Justus van Liebig, a famous chemist of the 1800s, this "law" originally referred to Liebig's theory that the yield of an agricultural crop is proportional to the amount of the most limiting nutrient. Among other versions or adaptations of this theory are the "law" that the capacity of a barrel with staves of unequal length is limited by the shortest stave, and that growth is controlled by the amount of the most limiting (often the scarcest) resource. Reference [104].
Limited License	*See* License Agreement.
Linch-Pin Technology	A technology that in some fashion holds a more complex structure together. For example, a linch-pin technology may be the key to commercializing certain other technologies, or it may transform the way other pre-existing technologies are used, or its deployment may require the invention or application of other technologies.
Linear Innovation Models	There are numerous models for the process of commercial innovation. The primary linear models are "*technology-push*" and "*demand-pull.*" These are both unidirectional. In *technology-push*, science drives discovery, which in turn drives invention, which in turn drives the development and commercialization

of products, process, or services into the marketplace. Since, it is *technology-push*, most or all of this process is conducted without regard for the existence of market niche or need, so the ultimate product, process, or service may not be commercially successful. Examples include the mass-market personal computer (1977: Commodore PET™, Apple™, and Tandy TRS-80™), Sony Walkman™ portable audio cassette player (1979), and the Apple iPad™ tablet computer (2010). Technology-push is also termed "*Discovery-Push*," "*Science-Push*," "*Supply-Side Driven*," or (in the older literature) "*Knowledge-Induced*." It is also termed the first, or the first-generation, model of innovation. In *market-pull*, marketplace opportunities drive the search for new product, process, or service concepts, for which innovators search the existing body of knowledge, inventions, and technologies. Since, it is *market-pull*, the market niche or need is clear, but the search for appropriate knowledge, invention, and/or product conceptualization and development may not be successful. Examples include digital photo editing software (1980s) and mass-market digital cameras (1990s). Market-pull is also termed *Demand-Pull, Needs-Pull*, or (in the older literature) *Demand-Induced*. It is also termed the second, or the second-generation model of innovation. *See* Figure 7. *See also* Non-Linear Innovation Models, Generations of Innovation.

Linear Thinking — *See* Creative Thinking Models.

Linked-Chain Model — *See* Chain-Linked Model.

Local Innovation — *See* Reverse Innovation.

Localist Regional Innovation System — *See* Regional Innovation System Types, Table 6.

Localization — *See* Internationalization (Product, Process, or Service).

Logistic Curve — *See* S-Curve.

Logistics Management — *See* Value Chain.

Long Cycles — *See* Kondratieff Waves.

Long Economic Cycles	*See* Kondratieff Waves.
Long Waves	*See* Kondratieff Waves.
"Looks Like" Model	*See* "Works Like" Model.
Low-Tech	Low-Technology. *See* High-Technology.
Ludd, Ned	*See* Luddite.
Luddite	Generally, a person that is strongly opposed to mechanization, industrialization, automation, or even new technologies of any kind. The original Luddites were textile workers that, fearing job losses, destroyed textile machines in England in the Industrial Revolution of the early 1800s. They are thought to have taken their name from Ned Ludd, a textile worker that had destroyed such a machine in 1779. Modern usage frequently refers to a person that is strongly opposed, or resistant, to the adoption of new technologies of any kind (hence "*Neo-Luddism*" and the *avoiders*, or *non-adopters*, in Technology Acceptance Models). The "Luddite fallacy" refers to the false hope that avoiding technological advances can in some way ensure economic success and/or sustainability. *See also* Technology Acceptance Model.
Luminaries	*See* Lead Users.

M

Maclaurin, (William) Rupert (1907–1959) A New Zealand-born, American economic historian known for his development of the first model of the process of innovation (building upon the industrial research cycle work of Maurice Holland). His "*theory of technological change*" laid out a sequential series of steps beginning with research and ending with commercialization that is now known as the technology-push linear model of innovation. He also suggested a number of measures that could be used as indicators of progress in each element of the sequence. It may have been Maclaurin that coined the term "*technological innovation*." Reference [105]. *See also* Holland.

Macroeconomics A sub-discipline of the field of economics that focuses on entire economies, as opposed to specific markets within an economy. Macroeconomics includes the structures, behaviours, and performance of economies, which may be regional, national, or broader. Macroeconomics is different from microeconomics, which focuses more narrowly on specific markets, although they are linked because the health of economies affect the actions of businesses and consumers and *vice versa*. *See* Microeconomics. *See also* Competitiveness, Gross Domestic Product, Kondratieff Waves.

Main Science and Technology Indicators (MSTI) Indicators of the degree and nature of science and technology activities of countries, particularly those assessed for this for this purpose by OECD. The OECD MSTIs cover selected inputs, such as research and development (R&D) spending and numbers of researchers, activities, such as R&D activities by sector; outputs, such as patents; outcomes, such as exports or export market share; and impacts, such as balances of trade or payments and international trade, jobs, and GDP growth. Reference [106].

Management Innovation New ways of organizing work or work-related functions, practices, or processes, including the management function, within an organization. Also termed "Managerial Innovation." *See also* Innovation, Organizational Innovation.

DOI 10.1515/9783110429176-014

Mandate Effectiveness Ratio (MER) An innovation metric pioneered by research and technology organizations (RTOs) and used to help quantify the amount of incremental economic activity that results from their work helping businesses address their technological challenges and opportunities in the marketplace. The definition is:

MER = (incremental economic impact)/(government investment in the RTO)

The timeframe is usually 1 year but should be specified. Also termed Mission Effectiveness Ratio. Reference [107]. *See also* Innovation Metrics.

Manufacturing Development *See* Research and Development.

Mark I Innovation *See* Schumpeter Mark I Innovation.

Mark II Innovation *See* Schumpeter Mark I Innovation.

Market Analysis In the commercialization of a product, process, or service, several elements have to be considered to understand the market potential. A typical sequence includes Market Identification, Market Size, Customers, Distribution Channels, and Competitors and Product Differentiation in the Marketplace. Reference [6].

Market Barrier *See* Innovation Barrier.

Market-Based Economy *See* Knowledge-Based Economy.

Market Demonstration *See* Demonstration.

Market Development Strategy *See* Ansoff Matrix.

Market Segmentation In terms of innovation strategy, a person or organization that takes advantage of a new idea, technology, invention, or technological innovation (developed by a "first mover") and implements a customized adaptation of it that is aimed at one or more niche markets. A "*Specialist*" strategy (also termed a "*Market Segmentation Strategy*") has a goal of serving niche markets with customized versions of the basic product, process, or service, usually later in the core technology's life-cycle trajectory. *See also* First Mover, Fast Follower, Late to Market.

Market Identification — A summary of the customer needs that will be satisfied, the specific customers who will buy the product, process, or service, the reason(s) they will buy it, and whether the timing is right in the marketplace. *See also* Market Analysis. Reference [6].

Marketing — The processes of identifying a company's main differentiating features, developing the key messages that explain these differentiators and their relevance, and developing and implementing strategies to position the company in its appropriate marketplace(s). Such strategies may include such things as branding, stakeholder relations, advertising, events, and communications (print, internet, and social media), for example. The principal components of marketing are *Market Research and Planning* and *Market Management*. Marketing helps to create visibility and profile (brand awareness) for a company, and sets the stage for Business Development and Sales. *See also* Business Development, Sales, Distribution, Innovation Barrier.

Marketing Innovation — New ways of marketing products, processes, or services that have have been successfully implemented and improve market penetration and/or market size. *See also* Innovation, Organizational Innovation.

Market Intelligence — *See* Competitive Intelligence.

Market Leader — *See* First Mover.

Market Management — This includes advertising, promotion, and customer service related to a product, process, or service. This forms part of *Marketing and Sales*. *See also* Market Research and Planning, Sales, Distribution.

Market Penetration Strategy — *See* Ansoff Matrix.

Market-Pull Innovation	*See* Linear Innovation Models, Incremental Innovation. *See also* Disruptive Innovation.
Market Readers	*See* Architects, Moonlighters.
Market Research and Planning	The evaluation of the market and the sales potential of a product, process, or service; market barriers; distribution channels; and identification of the target customers. This forms part of *Marketing and Sales*. *See also* Market Management, Sales, Distribution.
Market Size	A description of the target market and/or market segments for a product, process, or service; and also the size of those markets in numbers of units that can be sold. *See also* Market Analysis. Reference [6].
Mass Customization	*See* Product Platform.
Material Technology	A technology category comprising physical things, such as a tangible manufactured product. A process could be either a material or a social technology. *See also* Technology, Social Technology.
Material Testing Agreement	(MTA) A legal contract created to protect information considered to be proprietary or otherwise business confidential when materials are transferred from one party to another for testing, evaluation, or further development. The parties to the contract agree not to disclose to others the information covered by the agreement. Also termed *Material Transfer Agreement*, *Plant Material Testing Agreement (for plant materials, pMTA)*. *See also* Non-Disclosure Agreement, Option Agreement, Boilerplate.
Material Transfer Agreement	(MTA) *See* Material Testing Agreement.
McKinsey Global Institute Connectedness Index	*See* Connectedness Index.
ME	Medium-Sized Enterprise. *See* Small- and/or Medium-Sized Enterprise.

Mechanistic Model — A model for the innovation process in which innovation arises from an accumulation of numerous smaller steps or ideas that occur over a reasonably long period of time. *See also* Cumulative Synthesis Model, Transcendentalist Model.

Medici Effect — When exposure to a range of fields, disciplines, or cultures leads to a combination of existing concepts that sparks a creative new idea. The term "intersection" in the Medici Effect refers to the making of multiple connections among the various fields of knowledge, philosophies, and approaches. The name Medici refers to a family in fifteenth-century Italy that funded creators from a wide range of disciplines, drew them to a single location (Florence), and helped trigger the European Renaissance in such fields as art, science, politics, literature, and architecture. Reference [108]. *See also* Creative Thinking Models.

Medium-High-Tech — Medium-High-Technology. *See* High-Technology.

Medium-Low-Tech — Medium-Low-Technology. *See* High-Technology.

Medium-Sized Enterprise — (ME) *See* Small- and/or Medium-Sized Enterprise.

MER — *See* Mandate Effectiveness Ratio.

Messy Fireworks Innovation — *See* Non-Linear Innovation Models.

Mesotechnology — *See* Microtechnology.

Metrics — (Innovation) Measures used to quantify and compare innovation activities, outcomes, outputs, reach and/or impacts. These are usually distinct from, but may overlap with or supplement traditional business metrics such as production, sales, net income, and market share. Example: New patents applied-for and/or issued. For some specific examples see New Product Vitality Index, Mandate Effectiveness Ratio. *See also* Activities, Inputs, Outcomes, Outputs, Reach, Impacts.

MFP — *See* Multifactor Productivity.

MGI CI — McKinsey Global Institute Connectedness Index. *See* Connectedness Index.

Micro — A prefix used to signify one-millionth part of a unit, a multiple of 10^{-6}. The common symbol for micro is μ, as in μm for micrometer. *See also* Nano.

Microeconomics — A sub-discipline of the field of economics that focuses on specific markets within an economy, as opposed to entire economies. Microeconomics includes the positioning and behaviours of businesses and consumers in specific markets. Microeconomics is different from macroeconomics, which focuses on economies as a whole, although they are linked because the actions of businesses and consumers affect economies and *vice versa*. *See* Macroeconomics. *See also* Competitiveness, Gross Domestic Product.

Micro-Entity — *See* Small- and/or Medium-Sized Enterprise.

Micro-Sized Enterprise — *See* Small- and/or Medium-Sized Enterprise.

Micro-, Small-, and/or Medium-Sized Enterprise — *See* Small- and/or Medium-Sized Enterprise.

Microtechnology — A materials science term referring to materials or structures at a scale of about one micrometre (typically 0.1 to 100 μm), as opposed to the nanoscale (nanotechnology; 0.1 to 100 nm). The term "mesotechnology" has been used to describe the regime in-between these two, although others are still using this term as a synonym for microtechnoogy. *See also* Nanotechnology, Molecular Nanotechnology.

Mind Maps — *See* Spider Diagrams, *under* Creative Thinking Models.

Miners — A characterization of one of four kinds of organizational approaches to innovation strategy. "*Miners*" tend to be very large companies with lots of technology and process expertise, very large internal business units, and strong silos. They also tend to be weak on internal innovation, necessitating the use of external innovation resources and/or organizations, focused on finding innovation opportunities within the organization, and innovating through formal, centralized processes working through the external innovation organization. Jaruzelski and Dehoff have referred to such organizations as "*Technology Drivers*" [19]. *See* Innovation Strategy Mapping.

Minnesota Mining and Manufacturing Co.	*See* 3M Inc.
Mission Effectiveness Ratio	*See* Mandate Effectiveness Ratio.
Mission-Oriented Research	*See* Research and Development.
MNC	*See* Multinational Enterprise.
MNE	*See* Multinational Enterprise.
Mockup	A model built to accompany a *Product Definition* and used to illustrate a product or process, and sometimes also aspects of the environment in which it would be used. For example, a mockup for a process concept might include a model of some of the process vessels, pumps, and piping in which it would be used. Mockups are usually built to scale, but usually are not working models. *See also* Prototype. Reference [6].
Mode 1	*See* Modes of Science.
Mode 2	*See* Modes of Science.
Model	*See* Prototype.
Models of Innovation	*See* Linear Innovation Models, Non-Linear Innovation Models.
Modes of Science	A social science term for the way scientific knowledge is produced. Mode 1 refers to knowledge produced by discovery research conducted within specific disciplines (biology, chemistry, physics, etc.). Mode 2 refers to applied, mission-oriented research produced by multidisciplinary teams. Mode 2 research has also been connected to the "*entrepreneurial*

	university" and to university research in the Triple-Helix Model. Also termed Context-Driven Research, Post-Academic Science. *See* Research and Development. *See also* Entrepreneurial University, Triple-Helix Model.
Modular Innovation	*See* Architectural Innovation.
Molecular Engineering	*See* Nanotechnology.
Molecular Fabrication	*See* Nanotechnology.
Molecular Manufacturing	*See* Molecular Nanotechnology, Nanotechnology. *See also* Generations of Nanotechnology.
Molecular Nanotechnology	The organization of atoms and molecules from the nanoscale up, in order to produce structures and/or materials having precise, predictable properties. Synonyms include 'fourth-generation nanotechnology', molecular manufacturing, molecular engineering, and molecular fabrication. The nanoscale structures created are sometimes termed nanoscale architectures, or nanoarchitectures. *See also* Nanotechnology, Generations of Nanotechnology.
Moonlighters	A characterization of one of four kinds of organizational approaches to innovation strategy. "*Moonlighters*" tend to be large companies with lots of technology and process expertise, that look inward to develop innovations based on internal ideas and know-how. They also tend to be focused on a strong innovation culture that supports internal experimental and development project initiatives, including internal "moonlighting," willing to fund internally generated innovation projects, and innovating through bottom-up processes. Jaruzelski and Dehoff have referred to such organizations (and also to "Architects") as "*Market Readers*" [19]. *See* Innovation Strategy Mapping. *See also* Bottom-Up Innovation.
MSTI	*See* Main Science and Technology Indicators.
MTA	Material Testing (or Transfer) Agreement. *See* Material Testing Agreement.

Multifactor Productivity
(MFP) An industrial measure of the change in output per unit of a number of combined inputs (including labor, materials, and capital). Any change in MFP is due to factors other than the combined inputs, such as technology changes, change of scale efficiencies, production changes, management changes, and changes due to technological innovation). Although it is only one of these other factors, MFP is sometimes used as an indicator of the impact of changes due to technological innovation, and sometimes as an indicator of the amount of technological innovation. This could be for an industry or for an individual enterprise. Reference [109]. *See also* Total Factor Productivity.

Multinational Corporation
See Multinational Enterprise.

Multinational Enterprise
(MNE) A corporation that owns and/or controls facilities, assets and usually production, in at least one country beyond their home country. An MNE generally coordinates operations from a central (head) office. Some MNEs develop new products in one or more countries, then have them manufactured in other countries, and then have them distributed and sold in yet other countries, in order to optimize materials costs, labor, costs, and/or trade advantages. Also termed Multinational Corporation (MNC), Transnational Corporation, International Corporation, Stateless Corporation. In some usage, Global Multinational Enterprises (Global MNEs) are distinguished from regionally focused MNEs, the former having substantial sales in more than one or two continents and the latter having most of its sales in only one or two continents.

Multiparty Innovation
See Ecosystem Innovation.

Multiplier
See Impact Multiplier.

Multiple Discovery Theory
The hypothesis that most scientific discoveries and inventions are made essentially simultaneously by multiple (independent) discoverers and inventors, respectively. Also termed "*Multiple Invention Theory*" or "*Simultaneous Discovery (or Invention) Theory*." Commonly cited examples include the theory of the origin (evolution) of species and the formulation of calculus. Merton postulated that this could be the result of a

combination of the prerequisite knowledge and tools becoming available and having a substantial number of people pursuing similar investigations, possibly in response to similar social and intellectual forces [110]. Epstein points out that, in addition, most inventions are improvements of previous inventions [111]. An opposing theory is the "Heroic Theory of Discovery or Invention," which is the hypothesis that the principal discoverers and inventors of most discoveries and inventions, respectively, are rare "*greats*" or "*geniuses*."

Multiple Invention Theory — *See* Multiple Discovery Theory.

N

Nanoarchitecture — *See* Molecular Nanotechnology.

Nano — A prefix used to signify one-billionth part of a unit, a multiple of 10^{-9}. The common symbol for nano is n, as in nm for nanometer. *See also* Micro.

Nanomaterial — Any material having one or more nanoscale dimensions (0.1 to 100 nm). Use of this term is also usually taken to imply that the material has different properties than it would if not for the nanoscale dimensions.

Nanoscale Architectures — *See* Molecular Nanotechnology.

Nanotechnology — A rapidly growing area of materials science involving the design, characterization, manipulation, incorporation, and/or production of materials and structures in the nanoscale range (typically 0.1 to 100 nm), by any of a variety of physical and chemical methods. An important distinction is that these applications exploit the properties of the nanoscale components, distinct from bulk or macroscopic systems. There is a substantial overlap of scale between nanotechnology and microtechnology. Nanotechnology also encompasses the "nano" approach, or molecular nanotechnology, by which is meant the precise, controlled assembly of structures up from the molecular scale that are well-organized, and with reproducible properties. *See also* Microtechnology. Nanotechnology formed the basis for key developments in the 6th Kondratieff Wave of disruptive innovations and is one of the possible technological drivers of the 7th Wave of innovation. *See* 7th Wave, Kondratieff Waves.

National Innovation System — (NIS) *See* Innovation Ecosystem.

National System of Innovation — (NSI) *See* Innovation Ecosystem.

DOI 10.1515/9783110429176-015

NDA	*See* Non-Disclosure Agreement.
Need Seekers	*See* Explorers.
Needs-Pull Innovation	*See* Linear Innovation Models.
Negative Cash Flow	*See* Cash Flow.
Negative Innovation	The impacts of a failed innovation process that leaves an organization with reduced or less competitive product/process/service offerings in the marketplace than they had before commencing the innovation process. *See* Innovation Process Paradoxes (Innovation Process Paradoxes).
Neo-Classical Growth Model	*See* Solow-Swan Growth Model.
Neo-Luddism	*See* Luddite.
Net Perceptual Equity	*See* Perceptual Equity.
Net Rate of Replication	*See* Technological Capacity.
Network Regional Innovation System	*See* Regional Innovation System Types, Table 6.
New Growth Theory	*See* Endogenous Growth Theory.
New Media Age	*See* Technological Ages.
New Product Development Funnel	*See* Product Development Funnel.
New Product Development Process	(NPD) New product development processes generally involve a process for identifying new product ideas, followed by a process for deciding which and how many of them should be further explored or developed, and a process for re-evaluating after

each of a succession of development stages. An example of the sequence of stages is: identification of the problem or opportunity; investigation, idea generation, and solution identification; further investigation and proposal development; solution development and proof of concept; prototype testing, demonstration, and marketability testing; production, marketing, and commercial deployment; and post-project review. *See also* Stage-Gate® Product Development Process, Technology Stage-Gate Process, Waterfall Method.

New Product Sales Index — *See* New Product Vitality Index.

New Product Vitality Index — (NPVI) An innovation metric pioneered by 3M Inc. and used to help quantify the amount of a company's business and/or business growth originating from products that were introduced to the marketplace within a specific time frame (in 3M's case, the previous 5 years). The 3M definition is:

NPVI = (sales from products introduced within the past 5 years)/(total sales)

Also termed New Product Sales Index or Year 1 New Product Sales Index (where the time frame is one year). *See also* Innovation Metrics.

New-to-the-Firm Innovation — The adoption and implementation of a product, process, service, or function by an organization that had not previously done so. Example: adopting new management practices.

New-to-the-World Innovation — The first introduction of an innovation to a marketplace.

New-to-the-World Solution — *See* Breakthrough.

NGO — *See* Non-Government Organization.

Niche Fusion — A technology deployment strategy involving the introduction of a new product, process, or service into multiple market niches in order to achieve a greater competitive advantage in the combined niches than would have been possible with only a single niche. Reference [9]. *See also* Single Niche Strategy.

Niche Innovation — A form of innovation in which the new product, process, or service displaces or renders obsolete the competing products in a small market niche, but not an entire market. As such niche innovation falls between incremental and disruptive innovation. Reference [13]. *See* Figure 1. *See also* Disruptive Innovation; Evolutionary Innovation; Innovation; Incremental Innovation; Abernathy-Clark Model.

Niche Player — *See* Business Ecosystem.

NIH Syndrome — *See* Not Invented Here Syndrome.

NIS — National Innovation System. *See* Innovation Ecosystem.

Non-Commercial Innovation — *See* Innovation (Non-Commercial).

Non-Disclosure Agreement — (NDA) A legal contract created to protect information considered to be proprietary or otherwise business confidential. NDAs are used to enable parties to share confidential information with each other but otherwise maintain confidentiality. The parties to the contract agree not to disclose to others the information covered by the agreement. Also termed *Confidential Disclosure Agreement (CDA)*, *Confidentiality Agreement*. *See also* Material Testing Agreement. *See also* Public Disclosure, Boilerplate.

Nonexclusive License — *See* License Agreement.

Non-Government Organization — (NGO) A not-for-profit organization that is mostly or completely independent of government(s). The nature, purpose, means of revenue generation, and activities can very widely although a not-for-profit organization is driven by purpose ("Mission-driven") rather than the need to make profits and pay dividends to shareholders. Most NGOs have some kind of social and/or environmental purpose, and some are registered charities. The revenues for NGOs can be any combination of grants, donations, or revenue contracts, including fee-for-service contracts. Some NGOs perform contract services in the marketplace, whereas others are exclusively funded by donations. Not all NGOs are completely independent of governments however, as some NGOs receive substantial portions of their revenues in the form of government grants and therefore may subject to indirect influence and/or control by government(s).

Non-Linear Innovation Models There are a number of non-linear models for the process of how innovations are created, focused on the role of institutions in the process. The primary non-linear models are the "*coupling model,*" "*integrated model,*" and "*systems integration and networking model.*" The "*coupling model*" is sequential, drawing on technology-push and/or market-pull at each stage, but it is non-linear in that the various kinds of communication and interaction among contributors creates feedback loops. Over time the term coupling was gradually superseded by terms such as communication, interaction, and chain-linked. The term coupling has also been used in innovation with reference to the coupling of ideas in the minds of, for example, discoverers and inventors. The "*integrated innovation model*" involves parallel, rather than sequential development, with simultaneous R&D, prototype development, and manufacturing activities, for example. Integrated innovation processes encompass integrated development teams, engagement with suppliers and customers, and possibly engagement with partner organizations (horizontal collaboration). The "*system integration and networking model*" (SIN Innovation Model) involves parallel development integrating horizontal linkages with customers, suppliers, collaborators, marketers, and others throughout the process. The non-linear innovation models attempt to deal with the fact that technological innovation processes can be highly irregular, involving jagged lines of activity, and highly uncertain and they are sometimes referred to using metaphors such as "*fireworks*" or "*messy fireworks,*" hence the term "*Messy Fireworks Innovation.*" "*Innovation 2.0*" is a variation on this theme. *See* Figure 16. *See also* Linear Innovation Models, Generations of Innovation, Innovation 2.0, Innovation Ecosystem Models. References [112].

Non-Provisional Patent Application *See* Provisional Patent Application.

Figure 16: An Interactive Helix-Like Illustration of the SIN Innovation Model. Revolutions.

Non-Technological Innovation — Non-Commercial Innovation. *See* Innovation (Non-Commercial).

Not Invented Here Syndrome — (NIH Syndrome) A tendency of people, groups, or organizations to resist ideas, inventions, and/or innovations that have been developed by others. This frequently leads to rediscovering, reinventing, or redeveloping things that were already available elsewhere (often referred to as "*reinventing the wheel*"). The NIH Syndrome is generally attributed to some combination of belief that external groups and/or capabilities are inherently inferior, a desire for control and/or credit, or an undue attachment to the *status quo*.

Novelty Paradox — *See* Innovation Process Paradoxes (Innovation Process Paradoxes).

NPD — *See* New Product Development.

NPD Funnel — *See* Product Development Funnel.

NPVI — *See* New Product Vitality Index.

NSI — National System of Innovation or National Innovation System. *See* Innovation Ecosystem.

N-Tuple Helix Model — An innovation ecosystem (sociological) model developed by Etzkowitz and Leydesdorff, and others, as an expansion of the Triple-Helix and Quad-Helix Models, to include additional players, beyond governments, universities, intermediaries, and industry. References [113, 114]. *See also* Innovation Ecosystem Models, Triple-Helix Model, Quad-Helix Model.

O

OECD — Organisation for Economic Co-operation and Development, Paris, France. *See also* Frascati Manual, Oslo Manual.

OEM — *See* Original Equipment Manufacturer.

Open Information — Information that is openly available, without the need for purchasing, or licensing, or even significant interaction with the source of the information.

Open Innovation — The terms *Open Innovation* and *Closed Innovation* refer to two extremes in approaches a company can make to developing and commercializing technologies into innovations. In closed innovation, a company conducts essentially all of its innovation activities internally, often in a dedicated R&D Department, from idea generation through R&D and the subsequent activities such as experimental development, prototyping, pilot testing, demonstration, and preproduction. The main advantages of closed innovation are the ability to control the process and to potentially achieve dominating competitive advantage positions in the marketplace. The main disadvantages are cost and the fact that the company is limited by its internal idea generation and innovation capacity.

In open innovation, a company uses external resources to conduct some or even most of its innovation activities, especially in the areas of idea generation and "*adopt and adapt*" technology acquisition. The main advantages of open innovation are that a company is not limited by its internal idea generation and innovation capacity and that it may be less expensive to adopt and adapt ideas and technologies from others than to develop them internally. Another advantage of open innovation is that it lends itself to partnering with other organizations in the innovation ecosystem. The main disadvantages are that competitor companies can usually access the same external capabilities and may be able to exploit then in ways that are better, faster, and or cheaper making it difficult to achieve a competitive advantage in the marketplace. Although

DOI 10.1515/9783110429176-016

the foregoing represent extreme positions, many companies use a hybrid strategy and leverage internal innovation capabilities with external ideas, technologies, and other capabilities. For a variation on open innovations, see Permissionless Innovation. *See also* Collaborative Innovation, Innovation Ecosystem, Open Science, Inbound Open Innovation, Outbound Open Innovation, Open Source Innovation.

Open Science	The classical approach an organization or individual can make to conducting discovery research in science. In open science, scientists engage in a two-way exchange of ideas, data, and knowledge by collaborating with one another without charge, and by publicly disclosing (presenting and publishing) without charge and by applying only the restriction that users publicly acknowledge their sources of information and collaboration. This approach has some similarities with the concept of *Open Innovation*. Conversely, an example of Closed Science would be where a scientist makes use of publicly available ideas and information but then holds their own ideas, knowledge, and other results gained from their work confidential and does not release them publicly. This happens in some industrial R&D environments. *See also* Open Innovation.
Open Source Innovation	A kind of open innovation in which the process, source, and outcome of the innovation are open as opposed to private. Example: open source software. *See* Open Innovation.
Operating Expense	(OpEx) *See* Capital Expense.
OpEx	Operating Expense. *See* Capital Expense.
Opportunity Cost	The cost, usually meaning the loss of potential gain, that could be realized from alternatives when a particular opportunity or course of action is chosen instead.
Option Agreement	An agreement, often a legal contract, between two parties that grants one party the right to buy or obtain (or sell or dispose of) something at a specified price and at a specified future time. Example: An option agreement could be used to enable someone to try and/or evaluate a technology for a certain period of time before having to decide whether or not to buy

or license it. *See* Material Testing Agreement. *See also* Boilerplate.

Organizational Innovation — The ideation, development, and implementation of new and improved internal processes within an organization. Such new organizational processes should have some kind of efficiency or productivity benefit, even if they are internal and have little or no connection to commercialization or the marketplace. Also termed Institutional Innovation, although sometimes the term institution is meant more broadly. Examples: Administrative, Management, Marketing, Political, and Strategic Innovation. *See also* Innovation.

Organizational Paradox — *See* Innovation Process Paradoxes (Innovation Process Paradoxes).

Original Equipment Manufacturer — (OEM) A company that manufactures a part, or even a subsystem, that is integrated into another company's product. A similar part or subsystem made by someone else and not part of the original product is often referred to as "*aftermarket.*"

Originator — *See* First Mover.

Osborn-Parnes Model — A synonym for Creative Problem-Solving. *See* Creative Thinking Models.

Oslo Manual — An Organisation for Economic Co-operation and Development (OECD) document providing recommended methods for collecting and interpreting data on innovation. The OECD definition of innovation includes both *Commercial Innovation* and *Non-Commercial Innovation*. The counterpart of the Oslo Manual is the Frascati Manual, which is concerned with data on research and development. Reference [8]. *See also* Commercial Innovation, Frascati Manual.

Other Innovation Activities — Any or all of the scientific, engineering, technological, commercial, and financial steps – other than R&D – that are needed to achieve innovation. That is, the non-R&D steps needed to get products, processes, or services into the marketplace. Reference [8]. *See also* Oslo Manual.

Outbound Innovation	*See* Outbound Open Innovation.
Outbound Open Innovation	An aspect of the open innovation process in which the flow of technology or other information is from inside an organization to the outside. Example: licensing out of technologies. Also termed "Outbound Innovation." *See also* Open Innovation, Inbound Open Innovation.
Outcome-Driven Innovation	*See* Customer-Oriented Innovation.
Outcomes	(Innovation) The initial consequences of the things (outputs) that are produced by and/or happen as a result of activities conducted by an organization or its employees. Examples: The results of research and development activities. *See also* Activities, Inputs, Outputs, Reach, Metrics, Impacts.
Outliers	*See* Foresight.
Outputs	(Innovation) Things that are produced an organization or its employees. Examples: The communication, publication, and dissemination of the results of research and development activities, or the products, processes, and/or services produced and sold by a business. *See also* Activities, Inputs, Outcomes, Reach, Metrics, Impacts.

P

Pacemaker Technology — An early-stage technological innovation that has not yet been widely accepted and adopted by customers and whose applications, usefulness, and market potential are not yet well known. In some usage, the term is applied to the specific case for which the emerging technology is judged to have high potential for future applications. *See also* Key Technology; Incremental Innovation; Evolutionary Innovation; Disruptive Innovation.

PACE Process — *See* Product and Cycle-Time Excellence Process.

Paid-Up License — *See* Royalty-Free License.

P&ID — *See* Piping and Instrumentation Diagram.

Paradoxes of Innovation — *See* Innovation Paradoxes.

Parallel Thinking — *See* Creative Thinking Models.

Participative Innovation — In *Closed Innovation,* this refers to engaging others within an organization, beyond those officially in R&D, in innovation processes. Such organizations may appoint a Chief Innovation Officer, or the like, to lead and coordinate such processes on behalf of the organization. *See also* Open Innovation.

In *Open Innovation*, this refers to engaging others – both inside and outside of an organization, and beyond those officially in R&D, in innovation processes. Such organizations may appoint a Chief Innovation Officer, or the like, to lead and coordinate such processes on behalf of the organization. This meaning of the term *Participative Innovation* is essentially the same as Collaborative Innovation. *See also* Collaborative Innovation, Open Innovation.

Patent — A grant of property rights on new, useful, and non-obvious inventions by a government. The grant of a patent excludes,

DOI 10.1515/9783110429176-017

for a certain number of years, others from making, using, selling, offering to sell, or importing the invention in the specific country granting the patent. For this reason, many inventions are patented in multiple countries. Patents are generally awarded to the "first to file" a patent application rather than the "first to invent," although some countries allow a grace period before filing is required. For this reason, *provisional patent applications* are sometimes filed. The most common type patent is termed *Utility Patent*, which covers processes, machines, manufactured items, or composition of matter. Other types include the *Design Patent* (for designs related to manufactured items) and the *Plant Patent* (for certain kinds of new invented or discovered plants). *See* Invention, Intellectual Property, Patentable Invention, Provisional Patent Application. *See also* Prior Art.

Patentable Invention — *See* Invention, Patent.

Patent Applications — (Innovation Indicator) An indicator of the total number of practical inventions being protected in an economy is the number of patent applications registered per 10,000 personnel in the labour force. *See* Innovation Indicators and Tables 4 and 8.

Patent Database Mining — *See* Intellectual Property Portfolio Mining.

Patent Family — A group of patents that have been granted in multiple countries on the same invention. Example: OECD Triadic Patent Family, which is a group of patents on the same invention that have been granted by the European Patent Office (EPO), United States Patent and Trademark Office (USPTO), and the Japan Patent Office (JPO). The Triadic Patent Family concept was adopted by OECD in order to provide an indicator of technological change at the international level.

Patent Mining — *See* Intellectual Property Portfolio Mining.

Patent Pending — A term used to indicate that a patent application has been filed but not yet determined. That is, it has not yet been allowed, disallowed, or abandoned. *See* Patent.

Patent Protection

(Invention) Patent protection for an invention means that, once patented, an invention cannot be commercially made, used, or sold by anyone else without permission. *See* Patent.

(National System) Patent protection in a specific country refers to the degree of protection that a specific country provides for a patented invention and/or the relative strength of a country's intellectual property laws compared with others. *See* Ginarte-Park Index.

Pearl Curve

One of the simplest of the technology S-curve models, the Pearl logistic curve model relates the technology maturity (y) to time (t) as: $y = y_{\infty} / (1 + \alpha \cdot \exp^{-\beta t})$, where the maximum value for y is given by y_{∞}, and the shape of the S-curve is determined by the adjustable parameters α and β. *See also* S-Curve, Gompertz Curve.

Pencil

See "Does it Pencil?"

Per Capita

In innovation literature, the Latin term *per capita* is generally used to mean "per person." Example: GDP *per capita*.

Percentage of Patents with Foreign Co-Inventors

(Innovation Indicator) An indicator of the degree of connectedness and collaboration among a country's inventors and those in other countries is 100 times the ratio of the number of patents having foreign co-inventors to the total number of patents issued in a country. *See* Innovation Indicators and Table 4.

Perceptual Equity

The value of a brand, product, process, service, or enterprise as perceived by prospective customers or some other key stakeholder group. For example: in the case of customers, a high degree of perceptual equity for a particular product can contribute to a willingness to try the product, pay a premium price for it, and/or remain loyal to its brand versus those of competing products. *Net Perceptual Equity* refers to the difference between the positive perceptions about a brand or product and the negative such perceptions. *See also* Brand.

Perfectibility Hypothesis

(Innovation) The theory that a new technology in any particular field is perfectible, that continuing developments cause it to approach perfection relatively quickly leaving diminishing returns for future potential advances, and that as a result the technological maturity plateaus (i.e., the inventive potential becomes exhausted). This hypothesis has been used to attempt to explain technology S-curves. A counter-argument is provided by Schmookler (Reference [98]). *See* S-Curve.

Performance Trap — *See* Innovation Barrier.

Permissionless Innovation — A form of open innovation in which an organization can invite the public to conceive and submit new ideas, solutions to problems, analyses of data, and/or solutions to opportunities in return for some kind of incentive. Chesbrough and Van Alstyne refer to permissionless innovation "as a complement to traditional research and development." *See also* Open Innovation. Reference [115].

Pervasive Computing — The commonplace inclusion of computers in consumer products to enable them to communicate information to and from remote systems via the Internet. Also termed "Ubiquitous Computing." The devices themselves are sometimes termed IoT Devices (for Internet of Things Devices). An example is a Smart Meter replacement for a traditional electric power meter. *See also* Internet of Things.

PEST Analysis — Political, Economic, Social, and Technological Analysis. *See* Social, Technological, Economic, Environmental, and Political Analysis.

PESTEL Analysis — Political, Economic, Social, Technological, Environmental, and Legal Analysis. *See* Social, Technological, Economic, Environmental, and Political Analysis.

PESTLE Analysis — Political, Economic, Social, Technological, Legal, and Environmental Analysis. *See* Social, Technological, Economic, Environmental, and Political Analysis.

PFD — *See* Process Flow Diagram.

Phase 1–4 Trial — *See* Clinical Trial.

Phase–Gate Model — *See* Stage-Gate® Product Development Process.

Phase–Gate Process — *See* Stage-Gate® Product Development Process.

Pillars of Innovation — See Principles of Innovation.

Pilot — *See* Field Pilot.

Pilot Plant — *See* Field Pilot.

Pilot Test — *See* Field Pilot.

Pioneers — A category of potential customer or technology adopter in a psychological model of technology adoption. *See also* Technology Acceptance, Technology Acceptance Model, Technology Readiness Index, Technology Adoption Lifecycle.

Piping and Instrumentation Diagram — (P&ID) A diagram for a chemical manufacturing process that shows how the processing equipment and control instrumentation are interconnected. *See also* Field Pilot, Process Flow Diagram.

P-KIBS — Professional KIBS. *See* Knowledge Intensive Business Services.

Plant Breeders' Rights — A grant of exclusive commercial rights on the production and sale of the seeds of an original plant variety by a government. The plant variety has to be uniform, stable, and different from other varieties. Although commercial rights are protected, this does not extend to the production and/or growth of the variety for personal use. *See* Invention, Intellectual Property.

Plant Demonstration — *See* Demonstration.

Plant Patent — *See* Patent.

Plant Pilot — *See* Field Pilot.

Plant Test — *See* Field Pilot.

Plant Material Testing Agreement — (pMTA) *See* Material Testing Agreement.

Platform Innovation — When a new core technology, based on a technological breakthrough, leads to multiple new products, processes, or services. Example: The technological breakthrough of electronic imaging enabled an entire generation of digital camera products to be introduced whose core technology was very different from the previous celluloid film technology. It is possible to have a platform technology and create modular and/or architectural innovations within it, as long as the core technological breakthrough remains. Example: In digital data recording, magnetic tapes and a succession of smaller and smaller-sized floppy disks involved different materials and components but were all based on the same breakthrough core technology (magnetic-head recording). *See also* Henderson-Clark Model.

Platform Leader	*See* Business Ecosystem.
PLC	Product Life-Cycle. *See* S-Curve.
pMTA	Plant Material Testing Agreement. *See* Material Testing Agreement.
Polytechnic	An institute of technology that provides advanced education (and degrees) and/or vocational training (and diplomas). Some poytechnics also conduct advanced scientific and engineering research, and some conduct applied research.
Political Economy	*See* Knowledge-Based Economy.
Political Innovation	New government policies, programs, or legislation that help achieve social objectives. These could include improving service delivery or efficiency, or both. *See also* Innovation, Organizational Innovation.
Pontin's First Rule of Innovation	"Any sufficiently radical invention seems ridiculous to most people when they first encounter it" [116].
Pontin's Second Rule of Innovation	"The first attempt to commercialize an invention almost never succeeds" [116].
Porter, Michael Eugene (1947 – Present)	An American economist and business strategist best known to the innovation world for his work in business competitiveness at the industrial and national levels. Two of his most famous books are probably *Competitive Strategy* (1980) and *The Competitive Advantage of Nations* (1990). *See also* Reference [117].
Positive Cash Flow	*See* Cash Flow.

Post-Academic Science — *See* Modes of Science.

Post-Industrial Age — *See* Technological Ages.

Post-Modern Age — *See* Technological Ages.

Practice — *See* Technology.

Pragmatist — *See* Early Majority.

Pre-Venture Capital — Intermediate-stage, usually moderate-level capital for a commercialization process. (frequently termed *Start-Up Capital* for a new young company). The capital is typically raised through informal or formal investors, internal organizational funds, or government grants or loans. Pre-venture capital is sometimes enough to develop a technology from an engineering prototype or production prototype to the *Qualified Production Prototype* stage of product development, possibly even limited production and a first introduction to the marketplace. By this point, the market analysis and commercialization plan have probably evolved into a *Market Strategy* and *Business Plan. See also* Capital, Sweat Equity, Seed Capital, Start-Up Capital, Venture Capital.

Principles of Innovation — Any of a wide range of authors' selections (especially in the business and government policy literature) of key philosophies, attitudes, environments, or practices that are advocated as being conducive to encouraging, accelerating, or improving the success rates in innovation. There does not seem to be a single generally accepted set of such principles. Also termed *Pillars of Innovation, Tenets of Innovation. See also* Innovation, Research and Development, Activities, Other Innovation Activities, Ten Cs.

Prior Art — All of the publicly available knowledge related to an invention that existed in the world prior to the date that the invention was made (or, depending on a given country's patent laws, prior to the date of filing of a patent application on the invention). This includes the knowledge represented in previous patents, but also that contained in previous patent applications, publications, and sometimes even presentations. Prior art is one of the things assessed in judging the patentability of an invention. *See also* Invention, Patent.

Process	*See* Technology.
Process Benchmarking	*See* Benchmarking.
Process Flow Diagram	(PFD) A flowsheet diagram for a chemical manufacturing process that shows the principal processing equipment and the process flows that are intended to occur. *See also* Field Pilot, Piping, and Instrumentation Diagram.
Process Innovation	*See Innovation.*
Product	*See* Technology.
Product and Cycle-Time Excellence Process	(PACE Process) A gated process and management system for the development of new products, processes, or services, developed by Michael McGrath. Reference [118]. *See also* Idea-to-Launch Process, Stage-Gate® Product Development Process.
Product Champion	One of several kinds of stereotypical people that can play a critical role in the technological innovation process. A product champion is someone that can champion the innovation process throughout its stages of development with a clear goal of producing a successfully launched new product, process, or service in the marketplace. Regardless of the kind of organization, this person has to be able to guide the project through the "*valley of death*" and all other obstacles in its path. Having a product champion can greatly increase the success rate. *See also* Rainmaker, Leading-Edge Customer, Technological Gatekeeper.
Product Definition	A basic summary of a new product, process, or service including a description of what it is, a preliminary design, and a description of how it would be used. *See also* Business Concept, Prototype.
Product Development Funnel	Regardless of an organization's innovation development model, there will be a stage at which a number of potentially good ideas have been identified, but this number needs to be reduced in order to devote the most resources to the ideas with the best chance of being successful. Many organizations use the idea of a product development funnel, or hopper, in which many new ideas are fed in and there is a process for deciding which and

how many of them should be further explored or developed. This can be done over several stages, rather than just picking one from the funnel, and an organization may develop specific criteria for assessing the ideas at any given stage and deciding which ones to advance and which to either discard or move back to the front-end of the funnel. Ultimately, only a few, or even just one (in a small organization), may be selected for final development and commercialization. Also termed *Product Development Hopper, Product Development Pipeline, Innovation Funnel. See also* Stage-Gate® Product Development Process, Technology Stage-Gate Process.

Product Development Hopper — *See* Product Development Funnel.

Product Development Pipeline — *See* Product Development Funnel.

Product Development Process — *See* New Product Development Process.

Product Development Strategy — *See* Ansoff Matrix.

Product Diversification Strategy — *See* Ansoff Matrix.

Product Innovation — *See Innovation.*

Productivity Paradox — *See* Innovation Process Paradoxes (Innovation Process Paradoxes).

Production Prototype — A full-scale, completely operational, model of a product that has been built to mirror a mass-produced unit, but which has been custom built as a one-off or in very limited quantity. A production prototype is used to demonstrate performance requirements, including operation, safety, and durability. Also referred-to as a Design Model. *See also* Prototype. Reference [6].

Production Technology — *See* Technology.

Product Life-Cycle Curve — *See* S-Curve.

Product Platform — A group of components (hardware and/or software) that form the core of a range (family) of products. This enables a range of customized products to be developed for different markets and/or customer segments ("Mass Customization"). It also enables more sophisticated, higher-end products, termed "*Augmented Products*," to be built on the core product platform. Example: The printer-cartridge platform underlying Hewlett Packard ink-jet printers.

Product Qualification Prototype — *See* Qualified Production Prototype.

Product Technology — *See* Technology.

Professional Ladder — *See* Technical Ladder.

Profit Forecast — *See* Awareness, Trial, Availability, Repeat Model.

Pro Forma — A Latin phrase, which in the innovation world is usually used with the meaning "as a matter of form." It is used as an adjective or noun to describe a standardized type of document. For example, a "*pro forma* report," or a "*pro forma* financial statement." In the financial sector, "*pro forma* statements" often refer to corporate reporting that emphasizes current and/or projected figures, and as such may not comply with generally accepted accounting principles (GAAP).

Proof — *See* Demonstration, Proof of Concept, Prototype.

Proof of Concept — A physical demonstration that a technology concept can be made to work and is no longer a speculative concept. At this stage of development, the technology does not have to be practical, efficient, or cost effective. In terms of Technology Readiness Levels, a demonstrated proof of concept satisfies TRL level 3. *See* Figure 7. *See also* Speculative Concept, Technology Readiness Level.

Prosperity Gap — The difference in gross domestic product per capita (GDP per capita) between a country or region and some specified peer or peers. *See also* Gross Domestic Product.

Prototype — In the technical development of a product, process, or service, various kinds of prototypes, or models, are used for illustration, demonstration, and/or to provide technical data to aid in the process. A typical sequence includes a Mockup, Working Model, Engineering Prototype, Production Prototype, and Qualified Production Prototype. *See* Figure 7. References [5, 6]. *See also* Demonstration, Scale-Up, 'Works Like' Model.

Provisional Patent Application — A patent application that has been specifically filed as a "provisional application for patent" with the USPTO. Such an application allows filing relatively quickly and inexpensively, and without making specific patent claims, to establish an early filing date and to enable one to use the term "Patent Pending." However, a provisional patent application is only valid for 1 year, within which time one must either abandon the application or else file a full, non-provisional patent application. *See also* Invention, Intellectual Property, Patent, Patentable Invention.

Proxy Indicators — *See* Innovation Indicators.

Public Disclosure — Any communication of information or technology to a person or people that have not agreed to keep the information confidential. Under some laws, public disclosure must be in writing whereas under others it includes oral discussions or presentations, and/or "showing" or demonstrating a technology. Where patentability is concerned, public disclosure generally has to be substantial enough that it teaches or enables a party to duplicate an invention. *See also* Non-Disclosure Agreement.

Purchasing Unit — *See* Awareness, Trial, Availability, Repeat Model.

Pure Research — *See* Research and Development.

Q

Quad-Helix Model An innovation ecosystem (sociological) model developed by Etzkowitz and Leydesdorff and others as an expansion of the Triple-Helix Model to include intermediary organizations, hence: governments, universities, intermediaries, and industry. In the quad-helix model, intermediary organizations would be the primary incubators for many technology-based startups and most small- and medium-sized enterprises (SMEs). Also termed *Quadruple-Helix Model*. References [113, 119]. *See* Figure 12. *See also* Innovation Ecosystem Models, Innovation Ecosystem, Triple-Helix Model, N-Tuple Helix Model, Quad Model.

Quad Innovation Models The addition of a fourth kind of institution to any of the three-component innovation models can improve their sophistication and explanatory power. *See* Quad Model, Quad-Helix Model. *See* Figure 12. *See also* Innovation Ecosystem Models, Triple-Helix Model.

Quad Model An innovation ecosystem (sociological) model developed by Wilson in about 2003. The Quad Model is like a "Holy Trinity" Model to which a fourth segment has been added, comprising resourceful organizations that can act as catalysts, intermediaries, and/or entrepreneurs (hence the term "quad leaders"). The Quad Model's four kinds of institutions are: governments, research institutions (universities and RTOs), non-government organizations (NGOs), and industry. Another "*quad*" adaptation of the Holy Trinity Model is: governments, universities, intermediaries, and industry. References [119, 120]. *See also* Innovation Ecosystem Models, Innovation Ecosystem, Triple-Helix Model, Quad-Helix Model, N-Tuple Helix Model.

Quadruple-Helix Model *See* Quad-Helix Model.

Qualified Production Prototype A full-scale, completely operational sample of a product that has been built using a limited production run to demonstrate that it meets design standards. Qualified production prototypes

DOI 10.1515/9783110429176-018

are also used to demonstrate that they meet industry or regulatory standards. Also termed Product Qualification Prototype. *See also* Prototype. Reference [6].

Quick to Fail — *See* Fail Fast.

R

Radical Innovation — *See* Disruptive Innovation.

Radical Technological Transition — When the basic technology in a product, process, or service radically changes both the components in a product and the linkages between those components (i.e., the basic architecture). Examples: The technological change in ocean-going ships from sails to steam engines and the technological change in power shovels from steam to gasoline engines. *See also* Disruptive Innovation, Henderson-Clark Model.

Rainmaker — One of several kinds of stereotypical people that can play a critical role in the technological innovation process. A rainmaker is someone with highly developed creative abilities, who is good at problem identification and problem solving, and who can serve as a catalyst within a team. Also termed "Wizard." Such a person could be engaged at any point in a new product, process, or service development process (not just at the new idea and invention stages) and can greatly increase the success rate. The Rainmaker Index refers to a "creative potential" scale that is based on the Myers Briggs Type Indicator (MBTI) personality-type indicator system and that highly ranks people who score highly in the personality preferences intuitive (N) and thinking (T) in the MBTI system. *See also* Leading-Edge Customer, Product Champion, Technological Gatekeeper.

Rainmaker Index — *See* Rainmaker.

R&D — *See* Research and Development.

R&D Alliance — *See* Ecosystem Innovation.

R&D Expenditure — *See* Government R&D Expenditures, Gross Domestic Expenditure on R&D, Gross National Expenditure on R&D.

R&D Intensity — (Innovation Indicator) An indicator of the R&D investments from all sectors in an economy is the ratio of gross domestic

DOI 10.1515/9783110429176-019

	expenditure on R&D to gross domestic product (GERD/GDP). *See* Innovation Indicators and Table 4.
R&D Personnel	(Innovation Indicator) An indicator of the total number of people directly involved in R&D in an economy is the number of researchers involved in R&D per 10,000 personnel in the labor force. *See* Innovation Indicators and Tables 4 and 8.
R&D System	Research ansd Development System. *See* Innovation Ecosystem.
Rate of Replication	*See* Technological Capacity.
Rate of Return	*See* Return on Investment.
Rate of Technological Progress	*See* Technological Capacity.
Rate of Technology Dissemination	*See* Technological Capacity.
Rationalist Strategy	One of two traditional approaches to strategy and strategic planning: "rationalist" and "incremental." In essence, the rationalist approach consists of attempting to scan and understand the surrounding environment, develop a strategy taking the environmental scan into account, and implementing the strategy. This approach is similar to classical military strategy and builds upon the classical organizational SWOT (strengths, weaknesses, opportunities, and threats) approach to planning. In contrast, the incremental approach consists of developing and implementing tactics intended to take the organization in the desired general strategic direction, evaluate their effectiveness, and adjust and repeat as necessary. This kind of approach has also been referred to as "trial and error" strategy.
RCI	*See* Resource-Constrained Innovation.
Reach	(Innovation) The extent of individuals, groups, or markets that are made aware of and/or connected to product, process, service, or program outputs. Distinctions are sometimes drawn between potential customers for these outputs (i.e., primary

targets or primary customers) and potential partners, suppliers, or agents. *See also* Activities, Inputs, Outcomes, Outputs, Metrics, Impacts.

Recombinant Innovation	Innovation resulting from the adaptation of previous ideas or concepts. An example is when an idea is adopted from one product, process, service, sector, or country, then adapted and deployed in another. *See also* Adopt and Adapt.
Red Giants	*See* Innovation Performance Mapping.
Reduction to Practice	The translation of a new idea, discovery, or invention into a demonstrable practical application. Having achieved a reduction to practice is one of the differences between an invention and a patentable invention. *See also* Invention, Working Model.
Regional Innovation Cluster	*See* Cluster (Innovation Cluster), Innovative Regional Cluster.
Regional Innovation Hub	*See* Cluster (Innovation Cluster), Innovative Regional Cluster.
Regional Innovation System	*See* Innovation Ecosystem, Regional Innovation System Types.
Regional Innovation System Types	The character of regional innovation systems varies widely depending on the nature of the region and the depth of its pool of public and private innovation resources. A simplified taxonomy for regional innovation systems is illustrated in Table 6. Reference [121].
Re-Invention	(Business) Changing a company's business strategy, business model, products, or marketing approach in order to maintain or increase its competitiveness. (Customer) The degree to which an innovation is changed or modified by a user in the process of its adoption and implementation. Examples: A genetically modified seed would be difficult for a customer to modify, but a car is relatively easy for a customer to modify.

Research and Development

(R&D) Any or all of discovery research, applied research, and experimental development. *Discovery research* is experimental and/or theoretical investigation undertaken to acquire new knowledge and/or understanding of facts and/or phenomena – *without* any particular use or application in mind. Synonyms for discovery research include *Basic Research, Fundamental Research, Pure Research, Upstream Research, Upstream Studies. Applied Research* is experimental and/or theoretical investigation undertaken to acquire new knowledge and or understanding of facts and/or phenomena – but directed *with* a specific use or application in mind. Synonyms for applied research include *Mission-Oriented Research, Strategic Research. Experimental Development* is systematic work, drawing on existing knowledge gained from research and/or practical experience, aimed at producing new or improved materials, products, processes, systems, or services. Synonyms for experimental development include *Advanced Technology Development, Engineering Development, Manufacturing Development,* or simply *Development. See* Figure 7 and Table 7. *See also* Frascati Manual, Activities (Innovation), Modes of Science, Whole Product R&D. Reference [7].

Table 7: Technological Results Produced by Scientific and Engineering Research.

Technology-Producing Activities	Technological Results
Scientific Discovery Research	Knowledge and understanding of things and phenomena without regard to their utility.
Applied Scientific Research	Knowledge and understanding of useful and potentially useful things, phenomena, and processes.
Engineering Research	Knowledge and understanding of how to control things, phenomena, and processes.
Inventive Activities	Novel (invented) processes and products.
Development Engineering	Models and prototypes of novel, practical, and controlled processes and products.

Research and Development System

See Innovation Ecosystem.

Research and Technology Organization

(RTO) A government-owned corporation/agency or private not-for-profit company that is primarily focused on developing and deploying practical technologies that address commercial marketplace problems or opportunities, and spanning multiple sectors of an economy. RTOs differ from academia, mainstream government, for-profit companies, and even from mainstream not-for-profit companies. They have "public good" missions and are not primarily profit-driven. They work in the public interest but on market-pull issues, generally using a businesslike approach, and frequently with a secondary, socio-environmental agenda. One of the most important functions of an RTO is to help business enterprises access, absorb, adapt, deploy, and exploit new technologies in order to enhance businesses' ability to innovate and therefore their competitiveness and sustainability. Also termed Applied Research Organization. There are several associations of RTOs, including: World Association of Industrial and Technological Research Organizations (WAITRO), European Association of Research and Technology Organizations (EARTO), and Innoventures Canada Inc. (I-CAN). References [122, 123, 124].

Research and Technology Park — Real estate developments that provide infrastructure and a "like" environment for organizations with a strong interest in research, development, and/or commercialization of technology. Generally located adjacent, or in close proximity, to a major university Business Accelerator campus, research and technology parks generally comprise units of the local university itself, research and technology organizations (RTOs), industry, government agencies, small- and medium-size enterprises (SMEs), and even start-up companies. Some research and technology parks provide business services and even business incubation programs. Also termed Innovation Park, Research Park, Science Park, or Technology Park. *See also* Business Incubator.

Research Park — *See* Research and Technology Park.

Resource-Constrained Innovation — (RCI) Innovation (usually technological product innovation) developed in and for emerging economies in a context charac terized by financial and technological resource constraints. Also referred to as "*Innovation at the Bottom of the Pyramid.*" *See also* Reverse Innovation.

Return — *See* Return on Investment.

Return on Assets — *See* Return on Investment.

Return on Capital Employed	*See* Return on Investment.
Return on Equity	*See* Return on Investment.
Return on Innovation	(ROInn) A measure of the financial return on investment in innovation for one or more (usually specified) innovations, within a specified time period. Generally, a simple return on innovation would be the revenues derived from sales of a new product, process, or service less the costs associated with manufacturing, marketing, selling and shipping or providing, with the difference being divided by the costs associated with ideating, developing, and commercializing the innovation. Other measures could involve market results such as market share or new market penetration. Also termed "*Return on Innovation Investment*" (ROI2 or ROII). *See also* Return on Investment.
Return on Innovation Investment	(ROI2) *See* Return on Innovation.
Return on Investment	(ROI) An indicator of the effectiveness of an investment, given by the formula: ROI = (Gain from the investment − Cost of the investment) / (Cost of the investment) The ROI is usually expressed either as the simple quotient above or as a percentage. It is also sometimes used to compare the effectiveness of several different investments. Unless the investment is sold then the gain will refer to a specific time period, which should be clearly stated. For a given investment, the actual ROI value can be quite sensitive to the choices one makes in what to include in the gain and what to include in the costs, so the best practice is to include the complete basis for the calculation. ROI is sometimes termed *Holding Period Return*, *Rate of Return*, *Return*, or *Yield*. A common time period is 1 year, for which the ROI is sometimes termed the *Annual Return*. Finally, as an indicator of investment performance, ROI is different from return indicators based on organizational size or operations, such as *Return on Equity*, *Return on Assets*, or *Return on Capital Employed*. *See also* Return on Innovation.

Reverse Engineering — The process of determining how a product, process, or service works in order to design a competitive product. Example: By reverse engineering the IBM PC, competitors were able to develop and produce similar products that could be sold at lower prices. *See also* Fast Follower.

Reverse Innovation — The process by which innovations that have been developed to meet the needs of customers in developing nations are adapted (sometimes mostly by repackaging) for marketing and sales as low-cost innovations in developed nations. These are usually products rather than processes or services. The aspects developed in the developing nation(s) could be any or all of the product idea conception, the product research and development, the product design, or the initial market targeting. Example: GE's portable electronic medical instruments. Reverse innovation, in this sense, is as opposed to *Forward Innovation*, meaning the practice of starting with products that have been developed for developed nations and then adapting them (usually by removing features and costs) to produce innovations for marketing and sales in undeveloped nations. The first phase of reverse innovation is sometimes called Local Innovation. References [125,126,127]. Also termed *Blowback Innovation, Cost Innovation, Trickle-Up Innovation.* See also Innovation at the Bottom of the Pyramid.

Revolutionary Innovation — *See* Disruptive Innovation.

Richard's Curve — *See* S-Curve.

RIS — *See* Regional Innovation Systems, Regional Innovation System Types.

Rogers Diffusion of Innovation Model — *See* Rogers.

Rogers Diffusion of Technology Model — *See* Rogers.

Rogers, Everett M. (1931–2004) — An American sociologist and author best known in the innovation field for his theory and book on the diffusion of technologies in the marketplace (i.e., the diffusion of technological innovations). He used a normal distribution (i.e., bell curve) to describe the progress of technology diffusion by different adopters which he characterized as *innovators* (the first few adopters), *early adopters* (the next few adopters), *early majority, late majority*, and *laggards*. Rogers' "*Diffusion Of Innovation Model*" is sometimes referred to as the Innovation (or Technology) Diffusion Model or the Diffusion of Innovations (or Technologies) Model. Reference [31]. *See* Figure 6. *See also* Technology Adoption Lifecycle, Technology Dissemination, Technology Readiness.

ROI — *See* Return on Investment.

ROI2 — Return on Innovation Investment. *See* Return on Innovation.

ROII — Return on Innovation Investment. *See* Return on Innovation.

ROInn — *See* Return on Innovation.

Routine Innovation — *See* Subinvention.

Royalties — Payments made by a license for the right to use licensed intellectual property. *See* License Agreement, Royalty-Free License.

Royalty-Free License — An intellectual property License requiring no royalty payments. Such licenses are sometimes granted for free, sometimes as part of a bundle with another product or service, and sometimes following an up-front, one-time payment. Sometimes referred to as a *Paid-Up License*. *See also* Cross-License, License Agreement, Licensing, Royalties.

RTO — *See* Research and Technology Organization.

S

Sales — A company's sales function builds on the stage set by marketing and on the strategies developed by business development, to create exchanges of value for a company's products, processes, or services. The sales function helps identify, create, and maintain relationships with clients, customers, and/or users in order to maximize sales. Sales also includes the closing, or securing, of sales deals, which may be through exchanges of value or the signing of contracts. The process of identifying and nurturing customer relationships is usually conducted in partnership with the business development function. *See also* Marketing, Business Development, Distribution.

Sales and Distribution — This includes the sale process and sales force, and the distribution channels related to a product, process, or service. *See also* Market Management and Market Research and Planning.

Sales Forecast — *See* Awareness, Trial, Availability, Repeat Model.

Sandbox — Traditionally, this is a separate, usually multi-disciplinary, team in an organization that is charged with the task of exploring new ideas. In some cases, a short-list of ideas is handed off to other parts of the organization, while in others select new ideas are developed as far as the prototype stage. The term "sandbox" comes from the imagery of people coming together, in a specific place or environment, to "play" with new toys. The sandbox concept has also been applied to linkages among organizations by which people can interact and collaborate on new, pre-commercial, ideas for products, processes, or services. Also termed Innovation Sandbox. Another connotation of the term Innovation Sandbox is any idea generation and nurturing activity that involves free-form exploration and experimentation. Reference [128].

Saturation Curve — *See* S-Curve.

Scale-Up — Once a prospective new product or process has been shown to work at the laboratory-scale, or as a small-scale working

DOI 10.1515/9783110429176-020

model, a next step in its development is to design and test ("scale-up") a full-scale or intermediate-scale model. Depending on the nature of the product or process, a working model at larger scale might need to look and behave quite differently from the original. For example, in the case of a mechanical device some components might have to be made with different dimensions or from different materials in order to function properly at larger scale. In the case of a chemical process, the nature and geometry of the vessels, mixers, and other equipment might have to be quite different from their smaller-scale counterparts in order to properly scale-up critical process variables, such as shear and residence time, for example. *See also* Field Pilot, Prototype.

Sceptic *See* Laggards.

Schmookler, Jacob (1917–1967) An American economist known to the innovation world for his work studies of significant practical inventions and their adoption by industry. His work led him to identify (in 1966) the role of market demand ("*demand-pull*") as a driver for the invention process and what is now known as the "*market pull*" model[5] of technological innovation. He argued that although technological progress had been previously treated as an exogenous variable in economic models, it is actually primarily an endogenous variable and primarily driven by market demand. References [98, 129, 130]. *See also* Linear Innovation Models.

Schmookler Innovation In modern usage this refers to market-pull innovation, although Schmookler's actual work emphasized the role of economic conditions in driving the creation of practical, patented inventions. Schmookler's work covered the period from the industrial revolution to about 1950. A follow-up evaluation by Scherer in the 1970s, which covered more recent technological advances in industry, reached the same broad conclusion. *See* Market-Pull Innovation. References [98, 129].

Schumpeterian Growth Theory Schumpeterian Innovation. *See* Innovation (20th–21st century), Schumpeter Mark I Innovation.

---

5 The concept of market-pull seems to also have originated from the 1950s work of Carter and Williams [130].

Schumpeterian Innovation	*See* Innovation (20th–21st century), Schumpeter Mark I Innovation.
Schumpeter Innovation	*See* Innovation (20th–21st century), Schumpeter Mark I Innovation.
Schumpeter, Joseph Alois (1883–1950)	An Austrian economist and political scientist who is possibly best known for his theory of creative destruction as a necessary component of sustainable and/or growing economies. He also coined and/or championed the concepts of entrepreneurship and technological innovation as the agents of creative destruction. It also seems to have been Schumpeter that coined the term Kondratieff Waves. Reference [3, 4]. *See also* Kondratieff Waves.
Schumpeter Mark I Innovation	Schumpeter's original description of economic growth involved "creative destruction" in the marketplace driven by the introduction into the marketplace of "game-changing" new products, processes, or services by entrepreneurs or inherently entrepreneurial companies. The creative destruction involved the innovative companies succeeding over and/or replacing previous dominant companies that did not innovate. This has been termed Schumpeter Mark I innovation. In later work Schumpeter recognized another mode of innovation, in which dominating companies could maintain their competitive position by themselves introducing "game-changing" new products, processes, or services into the marketplace. This has been termed Schumpeter Mark II innovation. Such innovation-based activities by already dominating (usually mature and large) companies has sometimes been referred to as "creative accumulation," or "creative agglomeration," referring to the maintenance of competitive position by building an evolving and (usually) broadening portfolio of innovative product, processes, or services. From a broad market perspective, the Mark I and II innovation pathways are usually viewed as being complementary. *See also* Innovation (20th–21st century). References [3, 4].
Schumpeter Mark II Innovation	*See* Schumpeter Mark I Innovation.
Science Park	*See* Research and Technology Park.

Science-Push Innovation — *See* Linear Innovation Models.

Scientific and Engineering Research — Scientific research is the systematic investigation of hypotheses and theories (proposed explanations of things based on available facts and understandings) with the goal of generating new knowledge and understanding of things and phenomena. Engineering research is conducted somewhat similarly, but with the goal of generating new knowledge and understanding of how to make and/or control things. Some descriptions of technological results from scientific and engineering research are given in Table 7. *See also* Research and Development.

Table 8: Examples of Innovation Indicators for Organizations. References [156, 161, 162]

	Explanation
Lagging Indicators:	
Financial Returns from Innovation	Revenues and/or profits from the sale of new products, processes, or services.
Productivity Gains from Innovation	Productivity gains from the internal implementation of new products, processes, services, or technologies.
Market Share Gains from Innovation	Market share gains from the sale of new products, processes, or services.
Leading Indicators:	
R&D Spending	An indicator of an organization's investment in R&D.
Technology Investments	Investments in machinery, equipment and advanced technology, forming an indicator of an organization's investment in R&D.
Researchers	An indicator of the total number of people directly involved in R&D in the organization.
Patent Applications	An indicator of the total number of practical inventions being developed and protected by an organization.
External Patent Applications	An indicator of the total number of "significant" practical inventions being developed and protected (i.e., significant enough that they are worth protecting in other countries beyond that in which the inventions were made).
Triadic Patents Issued	Triadic patent families are groups of patents that have been granted in multiple (3 or more) countries on the same invention.
Trademarks	An indicator of the total number of new products and services being protected by an organization.
Existence of Internal Innovation Champion(s)	An individual who brings internal and external players and processes together to drive business processes needed to identify and evaluate opportunities.

Table 8 (continued)

	Explanation
Customer, Supplier, and Partner Engagement for New Ideas	Leading innovators bring these stakeholders "in" to help identify and evaluate opportunities.
External Partnerships and Collaborations	External partnerships and collaborations with customers and suppliers, as an indicator of an organization's level of collaboration with outside parties.
Formalized Business Development Processes	Leading innovators use formalized business processes and practices to help identify and evaluate opportunities.
Global Reach	Extent to which global activities are used to source new ideas and opportunities, not just markets for exported goods and/or services.

Scientific Revolution

See Technological Ages.

Scientometrics

The use of bibliometrics, that is, the study of the published literature, to measure the quality and impacts of scientific research.

S-Curve

Science. The first technological application of the S-curve concept appears to have been in 1845 by Pierre François Verhulst, who used it to illustrate stages in the development of yeast fungi colonies: a phase of initial growth and achievement of critical mass, a phase of rapid (approximately exponential) growth, a plateau-like phase of maturity, and a phase of decline due to depletion of the system's resources. Since then S-curves have been used to describe many other kinds of phenomena, sometimes involving four phases as in the example just given and sometimes with only the first three phases (i.e., without the decline phase). Synonyms include Growth Curve, Logistic Curve, Saturation Curve, Sigmoid Curve, Verhulst Curve. *See also* Pearl Curve, Gompertz Curve.

Technology. With a slight re-statement of the phases, S-curves can be used to describe the development and life-cycles of new technologies. The technology S-curve concept was introduced by Richard Foster in 1986 to help R&D managers. It involves a linear model of technology development, with overlays to illustrate how technologies get displaced by new ones. The typical phases are: (1) a phase of hypotheses, experimentation, addressing fundamental issues, and trial and error; (2) a phase of rapid maturing of the new technology as the fundamental issues become resolved, new approach(es) begin to take shape, individual advances begin to cluster, and/or a breakthrough occurs; and (3) a plateau as the technology matures and physical limitations are reached. The vertical axis of a technology S-curve is usually technology maturity and/or performance, while the horizontal axis is usually some kind of measure of cumulative research and development effort over time, summarized as R&D effort, investment, or time (*See* Figure 10). Synonyms include Saturation Curve, Sigmoid Curve, Innovation Continuum, Technology Growth Curve, Technology Life-Cycle, Technology S-Curve, Foster's Curve, Blindside Curve, Richard's Curve. *See also* Pearl Curve, Gompertz Curve, Technology Readiness. References [131, 132, 133].

Innovation. With another re-statement of the phases, S-curves can be used to describe degrees of innovation and their impact on an economy. Figure 11 provides an illustration comparing incremental, evolutionary, and disruptive innovation. Also termed Innovation Continuum (Degree of Innovation).

Technology Diffusion. A way of mapping the progress of technology diffusion in the marketplace. If technology adoption statistics are plotted cumulatively then an S-curve usually results. Such a representation demonstrates the plateauing of technology adoption by the time a technology is adopted by what Rogers termed the laggards. *See also* Technology Adoption Lifecycle, Technology Readiness.

Product, Process, or Service Lifecycle. Again, with a slight re-statement of the phases, S-curves can be used to describe the lifecycle of a product, process, or service. For a new product, the typical phases are (1) the introduction of a new product to the marketplace, with associated business development and early sales; (2) a phase of rapid sales growth as the product finds a successful market niche; (3) a sales plateau as the market niche becomes saturated; and (4) a phase of declining sales as the market-pull declines and/or competing products displace it in the marketplace. Each of the above phases can be considered in terms of factors supporting the product's development and factors acting against the product's development. Synonyms include Saturation Curve, Sigmoid Curve, Logistic Curve, Foster's Curve, Blindside Curve, Product Lifecycle (PLC) Curve. References [131, 132, 133]. *See also* J-Curve, Parl Curve, Technology Hype-Cycle, Buying Hierarchy.

In both the technology and product cases, the displacement of older technologies by newer ones, or of older products by newer ones, can be represented by families of S-curves. There is generally a gap, or transition phase, called a discontinuity (or technological discontinuity) between the plateauing of one S-curve and the rapid rise of a successive S-curve. Following this period of discontinuity the newer technology or product has overtaken the earlier one. Within a given type of product, process, or service a family of two or more S-curves, taken together is sometimes referred to as an Envelope S-curve. *See* Figure 17.

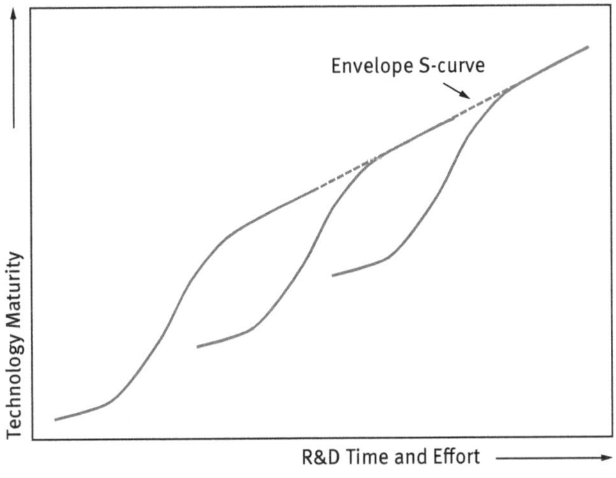

Figure 17: Illustration of a Family of Technology S-Curves.

Second Academic Revolution *See* Second Mission, Entrepreneurial University.

Second Generation Innovation Model The Market-Pull Model. *See* Generations of Innovation, Linear Innovation Models.

Second-Generation Nanotechnology	Molecular nanotechnology. *See* Generations of Nanotechnology.
Second Generation of Technology Foresight	*See* Generations of Technology Foresight.
Second Industrial Revolution	*See* Technological Ages.
Second Mission	The second mission of a university is research. The first mission is teaching. The evolution of universities to have a second mission has been referred to as the result of the "first academic revolution" from the teaching university. The third mission of an "*Entrepreneurial University*" involves participation by universities in working with governments and industry to enable economic growth (via innovation) and social progress. *See also* Entrepreneurial University, Modes of Science, Innovation Ecosystem Models, Triple-Helix Model, Quad-Helix Model.
Second to Market	*See* Fast Follower.
Second Wave	*See* 2nd Wave.
Seed Capital	Early-stage, usually modest capital for a company and/or commercialization process that is typically raised through informal investors, internal organizational funds, or government grants. In some cases, it comes from "*Angel Investors*" ("*Business Angels*"), family, friends, and entrepreneurs (sometimes termed "*Friends, Family, and Fools,*" or FFF Capital). Seed capital is sometimes enough to develop a technology from a mockup or working model stage to the Engineering Prototype or Production Prototype stage of product development and to develop a formal market analysis and a formal commercialization plan. *See also* Capital, Sweat Equity, Pre-Venture Capital, Venture Capital.
Self-Determined Innovation	*See* Indigenous Innovation.
Selling, General, and Administrative Expenditures	(SG&A Expenditures). *See* Capital Expense.

Sell-Off	A synonym for Spin-Off. *See* Spin-Out.
Service Innovation	*See* Innovation.
Service Mark	*See* Trademark.
Seventh Wave	*See* 7th Wave.
SG&A Expenditures	Selling, General, and Administrative Expenditures. *See* Capital Expense.
Shanzhai Innovation	A reference to the development of technological products, such as electronics, for which the appearance and/or the internal designs are imitative of, or even copies of, another company's products. The new products may or may not be of lower quality or lower price than the ones being adapted, imitated, or copied. Also termed "*Imitative Innovation.*" *See also* Indigenous Innovation.
Shooting Stars	*See* Innovation Performance Mapping.
Short-Term Mindset Trap	*See* Innovation Barrier.
Sigmoid Curve	*See* S-Curve.
Simultaneous Discovery (or Invention) Theory	*See* Multiple Discovery Theory.
Single Niche Strategy	A technology deployment strategy involving the introduction of a new product, process, or service into a single market niche in order to achieve a greater competitive advantage than would have been possible if introduced into multiple market niches. Reference [9]. *See also* Niche Fusion.
SIN Innovation Model	Systems Integration and Networking Model of Innovation. *See* Non-Linear Innovation Models. *See also* Generations of Innovation.
Singularity	The point in time at which truly intelligent machines come into being. The original concept of Vernor Vinge referred more specifically to the point in time at which machines with greater

than human intelligence come into being. Also called the Technological Singularity. Reference [134]. *See also* Deep Learning.

Sixth Wave — *See* 6th Wave.

Skeptics — A category of potential customer or technology adopter in a psychological model of technology adoption. *See also* Technology Acceptance, Technology Acceptance Model, Technology Readiness Index, Technology Adoption Lifecycle.

Skunkworks — Generally, the workings of a small team of people, often within an organization's research and development department, that are fairly unconstrained, unstructured, and aimed at testing and developing radical ideas. Some skunkworks are officially supported and may be well funded, whereas others are completely unofficial and may not have any explicit funding, and in either case they may be highly secret. Example: Lockheed Aircraft Corporation's World War II "*Skunk Works*" project that developed the XP-80 Shooting Star jet fighter in 1943.

SLEPT Analysis — Social, Legal, Economic, Political, and Technological Analysis. *See* Social, Technological, Economic, Environmental and Political Analysis.

Small- and/or Medium-Size Business — (SMB) *See* Small- and/or Medium-Sized Enterprise.

Small- and/or Medium-Sized Enterprise — (SME) Business enterprises that are smaller than a specified number of employees (e.g., 500), and/or have annual revenues of less than a specified value (e.g., $75 million), as distinguished from large-size enterprises. Some jurisdictions further distinguish between medium-, small-, and even micro-sized enterprises based on specific employee and/or revenue numbers. Example: In some usage, a Micro-Enterprise (or Micro-Entity) is one having up to 10 employees. Due to a combination of their large numbers and typically entrepreneurial natures SMEs are often referred to in discussions of innovation and economic growth. Sometimes referred to as *Micro-, Small-, and/or Medium-Sized Enterprise*, or *Small- and/or Medium-Sized Business (SMB).*

Small-Sized Enterprise — (SME) *See* Small- and/or Medium-Sized Enterprise.

Smart Cities — *See* Internet of Things.

Smart Grid — *See* Internet of Things.

Smart Manufacturing Age — *See* Technological Ages.

Smart Materials — Smart materials are those that sense and react to any of mechanical, thermal, chemical, electrical, and/or magnetic stimuli. Smart materials are sometimes subdivided into passive smart materials (sensors; sensing only), active smart materials (actuators; sensing and reacting), and very smart materials (sensing, reacting, and adapting). Examples include switchable surfactants, smart dyes, smart textiles, and self-healing materials. Also termed active materials. The related term intelligent materials usually refers to smart materials that also have the ability to self-control or self-regulate. *See also* Self-Healing Materials; Smart Dyes; Smart Textiles. *See* reference [135].

SMB — Small- and Medium-Size Business. *See* Small- and Medium-Size Enterprise.

SmE — Small-Sized Enterprise. *See* Small- and/or Medium-Sized Enterprise.

SME — *See* Small- and/or Medium-Sized Enterprise.

Social Innovation — The ideation, development, and deployment of new and improved solutions, processes, or practices within society in ways that have no connection to commercialization or the marketplace. These could take place in such areas as education, healthcare, or social services for example. A more accurate term would be *Social Invention*. *See also* Innovation.

Social Invention — *See* Social Innovation.

Social Internal Rate of Return on R&D — *See* Social Return on Investment on R&D.

Social Rate of Return on R&D — *See* Social Return on Investment on R&D.

Social Return on Investment in R&D — (SROI on R&D) An indicator of the net benefits to society of investment in applied research and development. The SROI on R&D is usually expressed as the increased value produced for society associated with the investment divided by the amount invested. The value component needs to be specified; it can comprise such things as social value, economic value, environmental value, or a combination of these. In the case of social or environmental value, some kind of financial proxy is usually developed in order to be able to complete the calculation. Example: One form of SROI on R&D is net incremental industrial productivity achieved within a specific time period divided by investment in applied R&D over the same time period. Also termed *Social Rate of Return on R&D, Social Internal Rate of Return on R&D.*

Social, Technological, Economic, Environmental and Political Analysis — (STEEP Analysis) A summary of environments and factors that could be assessed and considered as part of making business decisions, including those regarding technology commercialization and product, process, or service opportunities. Additional factors that are sometimes considered include legal, ethical, demographic, ecological, and/or regulatory. This has led to a host of acronyms including LEPEST, PEST, PESTEL, PESTLE, SLEPT, STEEPLE, STEEPLED, STEER and STEP. *See also* Competitive Intelligence, STEEPV Analysis.

Social, Technological, Economic, Ecological, Political, and Values-Based Analysis — *See* STEEPV Analysis.

Social Technology — A technology category comprising practical knowledge, such as the knowledge of how to construct something. Such knowledge can be intangible, such as knowledge that has to be transferred by demonstration and observation (tacit knowledge), or tangible, such as any form of codified knowledge. Witnin tacit knowledge are sometimes distinguished "*technical tacit knowledge*" (how to do things) and "*cognitive tacit knowledge*" (facts, structures, models, and beliefs) [9]. A process could be either a material or a social technology. *See also* Knowledge, Technology, Material Technology.

Socio-Institutional Innovation The ideation, development, and deployment of new and improved internal processes within a community organization. Such new organizational processes should have some kind of efficiency or productivity benefit, even if they are internal and have little or no connection to technological innovation. Example: producer co-operatives. This is a form of non-commercial innovation. *See* Innovation (Non-Commercial). Reference [33].

Soft Innovation Innovation that involves form, i.e., esthetics, rather than function. This is a legitimate form of innovation as long as it results in a commercially successful product, process, or service.

Solow Computer Paradox The risk, or "paradox," that increased investment in information technology could cause labor productivity to decrease instead of increase. In the early 1970s many organizations began investing in computer systems with the expectation that they would enable labor productivity improvements (or even eliminate labor completely in some areas). However, although computers and information technology enabled many companies to achieve competitive advantage and market share, the desired labor productivity increases were largely unrealized, and in some cases decreases were experienced. This is an example of the "*Productivity Paradox,*" *see* Innovation Process Paradoxes (Innovation Process Paradoxes). *See also* Solow, Robert.

Solow Growth Model *See* Solow-Swan Growth Model.

Solow, Robert (Merton) (1924 – Present) An American economist known for his work in macroeconomics, and especially for his theory of economic growth (also termed the neoclassical growth model). In 1987, he won the Nobel Prize in Economic Sciences for his analysis of economic growth. Solow's model of economic growth is termed the "Solow Growth Model" or the "Solow-Swan Growth Model" (having been independently developed and published in the same year – 1956 – by both Solow and Trevor W. Swan). It is also referred to as the "Solow-Swan Neo-Classical Growth Model." The Solow-Swan model separates the contributions to economic growth into increases in inputs (labor and capital) and technological progress, and predicts that sustainable economic progress requires labor – enhancing technological progress in order to increase output without needing more labor or capital. In his 1957 paper, Solow calculated that about

	four-fifths of the growth in US output per worker was attributable to such technological progress. Reference [136]. *See also* references [137, 138].
Solow-Swan Growth Model	An economic growth model. The Solow-Swan model separates the contributions to economic growth into increases in inputs (labour and capital) and technological progress, and predicts that sustainable economic progress requires labor – enhancing technological progress, in order to increase output without needing more labor or capital. In his 1957 paper, Solow calculated that about four-fifths of the growth in US output per worker was attributable to such technological progress.[6] Also termed "Solow Growth Model", or "Solow-Swan Neo-Classical Growth Model." *See also* Solow, Endogenous Growth Theory. References [137, 138].
Solow-Swan Neo-Classical Growth Model	*See* Solow-Swan Growth Model.
Specialist Strategy	*See* Market Segmentation.
Spectator	*See* Late Majority.
Speculative Concept	During the process of invention, various concepts will be developed that could link knowledge and/or discoveries to some kind of application. Such speculative concepts may or may not be feasible or even physically possible. At this stage of development, the technology does not yet have to be shown to be feasible. In terms of Technology Readiness Levels, a documented speculative concept satisfies TRL level 2. *See also* Proof of Concept, Technology Readiness Level.
Speed to Market	*See* Time to Market.
Spider Diagrams	*See* Creative Thinking Models.
Spillover Effect	*See* Impact Multiplier.
Spin-Off	*See* Spin-Out.

6 Also termed "exogenous technological change."

Spin-Out	In a "spin-out," a company allows outside investors to acquire minority ownership of a segment of it business (such as a division, business unit, or subsidiary). Such a segment is usually not a part of the company's core business(es), but of sufficient strategic interest that it is not completely divested (in contrast to a spin-off). In a "spin-off," a segment of a company (such as a division, business unit, or subsidiary) is completely divested and becomes either a separate company or a part of a separate company. Synonyms for spin-off include Sell-Off, Split-Off, and Carve-Out.
Split-Off	A synonym for Spin-Off. *See* Spin-Out.
SROI on R&D	*See* Social Return on Investment in R&D.
Stage 1 Economy	*See* Competitiveness Drivers.
Stage 2 Economy	*See* Competitiveness Drivers.
Stage 3 Economy	*See* Competitiveness Drivers.
Stage-Gate® Product Development Process	A stage-gate® process is a project management tool by which a project progresses through a series of stages, each one of which is followed by a gate. Each gate triggers a review of accomplishments and decision as to whether the project may proceed to the next stage. The gate evaluations are normally completed with reference to a set of pre-determined success criteria and can involve many kinds of input. The tool can be used as an aid to objectively and consistently managing efficiency and risk. References [38, 39]. *See* Figure 9 for an illustration. Also termed: Phase-Gate Model, Phase-Gate Process, Technology Stage-Gate Process. *See also* New Product Development Process, Waterfall Method.
Start-Up Capital	Intermediate-stage, usually moderate-level capital for a new young company (frequently termed Pre-Venture Capital for a commercialization process). Sometimes termed "*A-Round Capital.*" The capital is typically raised through informal or formal investors, internal organizational funds, or government grants or loans. By this point there will be a *Market Strategy* and *Business Plan*, but most of the development costs still lie ahead. Two stages are sometimes distinguished. *Early Stage* (or Formative Stage) refers to companies beginning operations,

and probably having a product or service in testing or pilot production, but which are not yet at the stage of commercial manufacturing and sales. *Later Stage* refers to companies beginning commercial manufacturing and sales, but before any initial public offering (IPO). *See also* Capital, Sweat Equity, Seed Capital, Pre-Venture Capital, Venture Capital.

Start-Up Company — A new, or relatively new, company that is still in an early stage of development and may not yet have well-developed products, services, markets, or sales. For this reason, start-ups often need external financing. *See also* Business Accelerator, Business Incubator.

Stateless Corporation — *See* Multinational Enterprise.

STEEP Analysis — *See* Social, Technological, Economic, Environmental and Political Analysis.

STEEPLE Analysis — Social, Technological, Economic, Environmental, Political, Legal, and Ethical Analysis. *See* Social, Technological, Economic, Environmental, and Political Analysis.

STEEPLED Analysis — Social, Technological, Economic, Environmental, Political, Legal, Ethical, and Demographic Analysis. *See* Social, Technological, Economic, Environmental, and Political Analysis.

STEEPV Analysis — A foresight evaluation process that involves considering possible future scenarios in terms of six points of view. The acronym STEEPV refers to social, technological, economic, ecological, political, and values-based points of view, and the idea is to evaluate a given situation or scenario from each of these six perspectives, as an aid to planning. *See also* Social, Technological, Economic, Environmental, and Political (STEEP) Analysis, SWOT Analysis, Delphi Method.

STEER Analysis — Social, Technological, Economic, Ecological, and Regulatory Analysis. *See* Social, Technological, Economic, Environmental and Political Analysis.

STEP Analysis — Social, Technological, Economic, and Political Analysis. *See* Social, Technological, Economic, Environmental, and Political Analysis.

Strengths, Weaknesses, Opportunities, and Threats	See SWOT Analysis.
Strategic Innovation	A kind of *Organizational Innovation*, strategic innovation refers to developing and implementing strategies for creating major and/or disruptive technological innovations (products/processes/services) that are intended to significantly improve an organization's performance versus its competitors. Ideally, such new strategies would create new value for the organization and for its customers, but it could be either. Such organizational innovations are probably the least common because they usually involve dramatic changes in how an organization operates. Accordingly, such changes have to be led, or at least supported, by CEOs. Example: The strategy change at General Electric under CEO Jack Welch in the USA in the 1980s. Another example is provided by Amazon.com, whose business strategy changed the breadth of product availability, distribution, and promotion, compared with traditional bookstores, while also reducing overhead costs and prices. Also termed Business Model Innovation. References [139, 140]. *See also* Organizational Innovation, Innovation.
Strategic Research	*See* Research and Development.
Strategy Innovation	*See* Strategic Innovation.
Subinvention	An older term for an invention that would have been obvious to one skilled in the art. A subinvention is therefore not a patentable invention but nevertheless represents something new and potentially useful. Subinventions are often created by skilled practitioners in a field, as incremental improvements or adaptations of existing processes or things. This has also been referred to as "*routine innovation.*" [98]. *See also* Invention.
Success/Failure Paradox	*See* Innovation Paradoxes (Innovation System Paradoxes).
Success Paradox	*See* Innovation Paradoxes (Innovation System Paradoxes).
Supercycles	*See* Kondratieff Waves.

Superstructure Organization	*See* Intermediary Organization.
Supply Chain Management	*See* Value Chain.
Supply-Side Driven	*See* Linear Innovation Models.
Sustainability-Related Triple-Helix	*See* Triple-Helix Model.
Sustaining Innovation	*See* Evolutionary Innovation.
Swan, Trevor W	*See* Solow, Robert.
Sweat Equity	Any unpaid work that is contributed by a person, group, or organization to the commercialization process. Sweat equity is sometimes enough to develop a technology to the *Working Model* stage of product development and to develop a high-level market analysis and a high-level commercialization plan. *See also* Capital, Seed Capital, Pre-Venture Capital, Venture Capital.
SWOT Analysis	An evaluation process for assessing a real or imagined, current or future situation or scenario and its potential impacts. The acronym SWOT refers to strengths, weaknesses, opportunities, and threats, and the idea is to evaluate a given situation or scenario from each of these four perspectives, as an aid to planning. For example, when faced with a new possible market situation, an organization might assess its own strengths and weaknesses in the context of that situation and consider the relative merits of avoiding or pursuing the potential opportunity or defending or fleeing from the potential threat. *See also* STEEPV, Delphi Method.
Synectics	*See* Creative Thinking Models.
Synthetic Innovation	Technological innovation that involves using existing knowledge and/or ideas in new ways (i.e., without involving research).
System Integration and Networking Model of Innovation	*See* Non-Linear Innovation Models. *See also* Generations of Innovation.

T

Tacit Knowledge	*See* Knowledge, Social Technology.
TAM	*See* Technology Acceptance Model.
Tangible Knowledge	*See* Social Technology.
Technical Innovation	*See* Technological Innovation.
Technical Intelligence	*See* Competitive Intelligence.
Technical Ladder	A career development pathway, available in some organizations, by which professionals can remain in a professional/technical career path as an alternative to the traditional management path. The idea is to motivate and retain experienced professional/technical staff that do not want to go into management by offering them additional career stages that either offer theoretically equivalent status and rewards or at least increasing status and rewards. The existence of two career development pathways in an organization is sometimes referred to as having a "Dual Ladder." Also termed "Professional Ladder" or "Individual Contributor Ladder."
Technical Revolution	*See* Technological Ages.
Technical Tacit Knowledge	*See* Social Technology.
Technological Ages	Technological Ages (or Revolutions) represent periods of major technological, economic, and societal changes. At least four industrial revolutions have been broadly identified. *See also* Kondratieff Waves and Figure 15.

DOI 10.1515/9783110429176-021

The First Industrial Revolution (mid-1700s to mid-1800s). Some of the principal changes included transitions from wood to coal power and hand to machine production methods (particularly in the textile industry), the advent of chemical and iron manufacturing processes, then increasing use of steam power and the development of machine tools, and the transition to high-volume printing via the steam-powered printing press. The first Kondratieff Wave lies within this period.

The Second Industrial Revolution (mid-1800s to mid-1900s). Some of the principal changes included the transition from coal to oil power, the advent of electricity and the telegraph and telephone, mass production, the internal combustion engine, highways, and rapid transportation. Also known as the Technological Revolution. Most of the second and all of the third Kondratieff Waves lie within this period.

The Third Industrial Revolution and Post-Industrial Age (mid-1900s to present). Some of the principal changes have included transitions from analogue to digital technology (particularly digital computers and communications), the advent of the Internet, information technology, mass media, and globalization. Also known as the Digital Age, New Media Age, Information Age, or the Post-Modern Age. Most of the fourth and all of the fifth Kondratieff Waves lie within this period.

Several Post-Industrial Revolutions (Ages or Societies) have been identified, beginning with the Scientific Revolution (with key developments including electricity, chemistry, and the chemical industry) and the Technical Revolution (with key developments including automobiles, mass production, and the petrochemical industry), followed by the Information Age, the Knowledge Age, and the Creative Age. An *Information Society* creates and disseminates information. Although this was not new, the growth of information and communications technology (ICT) massively increased data production and dissemination, and the Internet has made global connectivity possible and rapid. However, the production and communication of information alone does not necessarily lead to knowledge creation. A *Knowledge Society* develops, processes, shares, and uses knowledge to improve economic, social, and/or environmental conditions. Knowledge societies are by nature also *Learning Societies*, and embrace the concept of lifelong learning. Reference [141]. A *Creative Society* comprises people that are developing their natural creative talents and energies, and combining the power of information, knowledge, and creativity into an important economic force. Ref [142].

The Fourth Industrial Revolution and Smart Manufacturing (future). Some of the principal changes being predicted for a future "*Age*" include Smart Manufacturing, in which supply-chains, logistics, production, and product lifecycle management are all interconnected in systems that are much more intelligent, adaptable, and dynamic that they are at present, with the ability to improve through self-optimization and autonomous decision-making. The sixth and seventh Kondratieff Waves are in the future.

Technological Capacity

The amount of technological knowledge in an organization, region, or country, divided by the size of the labor force, or the population. Also termed "*Intellectual Capital.*" The rate of growth of an organization's, region's, or nation's technological capacity depends on the rate at which new technology is produced (the "*rate of technological progress*") and the rate at which technology is disseminated (the so-called *rate of replication*). Since some new technologies are incremental additions whereas others replace earlier technologies that are thus made obsolete, it is sometimes appropriate to distinguish among the "*net rate of replication*" and the "*gross rate of replication.*" Reference [98]. *See also* Technology.

Technological Determinism

See Fear of Innovation.

Technological Discontinuity

See S-curve.

Technological Gatekeeper

One of several kinds of stereotypical people that can play a critical role in the technological innovation process. A technological gatekeeper is someone that, on the one hand, acts as a conduit to external knowledge by maintaining a watching brief on relevant parts of the science and technology world, staying on top of the literature and technical conferences, and being knowledgeable of and/or actually linked to key external science and technology individuals (experts). On the other hand, this person acts as the communicator of such information to others involved in the technological innovation process. Such a person could be engaged at any point in a new product, process, or service development process (not just at the new idea and invention stages), ideally throughout the entire process, and can greatly increase the success rate. *See also* Rainmaker, Product Champion, Leading-Edge Customer.

Technological Innovation — *See* Innovation (20th–21st century). *See also* Maclaurin.

Technological Product and Process Innovation — (TPP Innovation) A technological innovation involving a product or process. *See* Innovation (20th–21st century).

Technological Progressiveness — A simple lagging indicator of technological innovation performance for commercial enterprises or entire industries, by which their performance in terms of having introduced "important" new or improved products or processes into the marketplace within a certain period of time is rated as being high, medium, or low. Introduced by Maclaurin in 1955. References [105, 143].

Technological Revolutions — *See* Technological Ages.

Technological Sophistication — *See* High-Technology.

Technological Unemployment — *See* Fear of Innovation.

Technological Valley of Death — *See* Valley of Death.

Technologist — *See* Innovator.

Technology — A broad term representing the knowledge of how to effectively use a product, process, or service (*know-how*), how to conduct or control a manufacturing activity (*practice, process, or production technology*), or how to manufacture, use, or consume something (*product technology*). In this context, a *tool* can be considered to be a product. Each of these kinds of technology can be protected as intellectual property, and each can be commercialized. Technologies are sometimes distinguished as being material technology (physical things) or social technology (intangible knowledge) although some kinds of technology, such as a process, could be either. *See also* Acquisition of Technology, Technological Capacity.

Technology Acceptance Model — (TAM) A psychological model developed to assess and describe how the features of a technology influence people's views of that technology in terms of perceived usefulness and perceived ease of use. Such views influence the extent to which a technology will be used by customers, if at all. A related, complementary model is the Technology Readiness Index (TRI), and there is also an integrated Technology Readiness and Acceptance Model (TRAM) that combines the TAM and TRI models to provide a single means of categorizing customer readiness to accept a new technology. In terms of psychological readiness of customers to accept and/or embrace new technologies, some literature makes reference to market segmentation in this context, a simple version of which would be to distinguish among *explorers* (the first few adopters), *pioneers* (the next few adopters), *skeptics*, *hesitators*, and finally the *avoiders* (the non-adopters), This has some analogy to the categorizations in Rogers' Technology Adoption Lifecycle. *See also* Acceptance Threshold, Technology Readiness, Technology Readiness Index, Technology Readiness Level, Technology Adoption Lifecycle, Acceptance Threshold, Luddite.

Technology Acceptor Capacity — The ability of a company or industry to accept, adapt (if necessary), and adopt new technologies. Also termed "Absorptive Capacity."

Technology Adoption Lifecycle — The stereotypical pattern of *Technology Diffusion* in which a new product, process, service, or idea becomes adopted and spreads according to the sociological characteristics of different adopters. Rogers [31] originally used a normal distribution (i.e., bell curve) to describe these different adopters as *innovators* (the first few adopters), *early adopters* (the next few adopters), *early majority, late majority*, and *laggards*. *See* Figure 6.

See also Technology Dissemination. See also Technology Hype-Cycle, Technology Readiness.

Technology Balance of Payments	The net technology transactions (in purchasing power parity US dollars) per 10,000 personnel in the labor force. This represents the balance of sales versus purchases of technology (such as patents, licenses, designs, trademarks, and trade secrets) in an economy. See Innovation Indicators and Table 4.
Technology Commercialization	*See* Commercialization.
Technology Dependency Ratio	The number of non-resident patent applications divided by the number of resident patent applications.
Technology Deployment	The processes involved in getting a new technology in place and operating properly in its intended working environment. Such processes generally include construction and/or installation, configuration, testing and/or troubleshooting, and making any necessary final modifications. Also termed "Technology Implementation."
Technology Diffusion	*See* Technology Dissemination.
Technology Diffusion Model	*See* Rogers.
Technology Dissemination	The transfer, adoption, and use of technology by multiple users. Dissemination is the way in which technology (and therefore innovation) spreads from its developer and first implementer and/or its first customer, to different conustomers, industries, markets, and regions. Two modes of technology dissemination are distinguished: *Technology Transfer* and *Technology Diffusion. Technology Transfer* refers to the transfer of technology from a specific person or organization to another and occurs for initial and/or limited technology dissemination. *Technology Diffusion* refers to the transfer of technology from a person or organization to another where either party may be unaware of the identity of the other, and occurs when technology is becoming widely disseminated. *See also* Technology Adoption Lifecycle, Technology Readiness.

Technology Drivers — *See* Miners.

Technology Foresight — *See* Foresight.

Technology Gap — *See* Innovation Gap.

Technology Gap Theory — *See* Innovation Gap Theory.

Technology Growth Curve — *See* S-curve.

Technology Hatchery — *See* Business Incubator.

Technology Hype-Cycle — A product lifecycle model in which a new product, process, or service advances through an unrealistic peak in market expectations, then sales decline until improvements and/or greater realism in the market prevails, after which the product/process/service matures in the market and then completes its lifecycle. Also termed Gartner Hype Cycle. A chart or graph illustrating these stages is referred to as a Hype-Cycle Curve or Gartner Hype-Cycle Curve. *See* Figure 18. *See also* Technology Adoption Lifecycle, S-curve.

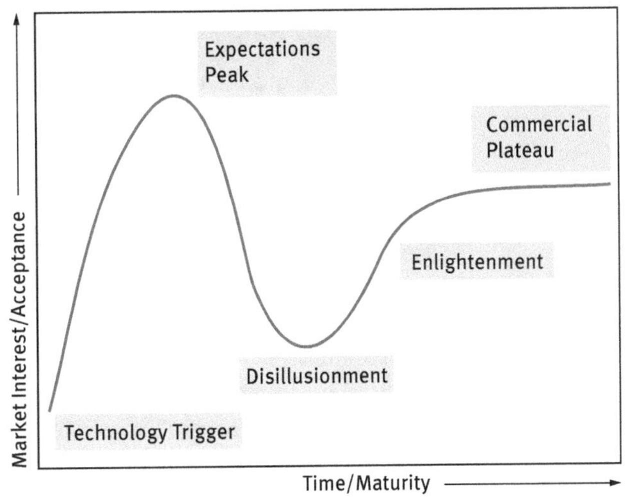

Figure 18: Illustration of a Technology Hype-Cycle Curve.

Technology Implementation	*See* Technology Deployment.
Technology Lifecycle	*See* S-curve.
Technology Management	An organizational function that coordinates efforts to manage the technologies involved in conducting the organization's functions. Such efforts could include technology strategy, technology foresight, technology road-mapping, technology portfolio management, and technology maintenance. Technology management is different from "*Innovation Management.*" *See also* Innovation Management.
Technology Market Intermediaries	(TMI) *See* Knowledge-Intensive Business Services, Intermediary Organization.
Technology Mudslide Hypothesis	A phrase coined by Clayton Christensen, referring to the difficulties companies face in trying to cope with the "relentless onslaught of technology change" in their respective markets as being like "*trying to climb a mudslide raging down a hill. You have to scramble with everything you've got to stay on top of it, and if you ever once stop to catch your breath, you get buried.*" This assumes that all of the competing companies in a market niche are constantly having to move "upward" in terms of innovations or else they will slide "downhill" and lose market share. Reference [14]. *See also* Incremental Innovation.
Technology Park	*See* Research and Technology Park.
Technology-Push Innovation	*See* Linear Innovation Models, Disruptive Innovation. *See also* Incremental Innovation.
Technology Readiness	There are two very different kinds of technology readiness: the readiness of a technology for deployment in the marketplace, and the readiness of customers to accept and/or pay for the new technology. The issues determining the former are primarily technical and are illustrated by, for example, the Technology Readiness Level (TRL) scale. The issues determining the latter are primarily psychological and are illustrated by, for example, the Technology Readiness Index (TRI) and the Technology Acceptance Model (TAM) scales. *See* Technology

Readiness Index, Technology Readiness Level, Technology Acceptance Model.

Technology Readiness Index — (TRI) A psychological model developed to assess and describe how people's personalities influence the degree of acceptance and use of a new technology. The TRI represents an attempt to balance factors that would contribute to technology acceptance (optimism about the technology and personal association with innovativeness), versus factors that would inhibit technology acceptance (discomfort caused by the technology and insecurity caused by mistrust of the technology). Such views influence the extent to which a technology will be used. The original TRI model comprises 36 elements, while a more recent "refined" TRI model comprises 16 elements. *See also* References [144, 145]. A related, complementary model is the Technology Acceptance Model (TAM), and there is also an integrated Technology Readiness and Acceptance Model (TRAM) that combines the TAM and TRI models to provide a single means of categorizing customer readiness to accept a new technology. The TRAM approach meshes the personality-related features of TRI with the system-related features of TAM. In terms of psychological readiness of customers to accept and/or embrace new technologies some literature makes reference to market segmentation in this context, a simple version of which would be to distinguish among *explorers* (the first few adopters), *pioneers* (the next few adopters), *skeptics*, *hesitators*, and finally the *avoiders* (the non-adopters), This has some analogy to the categorizations in Rogers' Technology Adoption Lifecycle. *See also* Acceptance Threshold, Technology Readiness, Technology Readiness Level, Technology Acceptance Model, Technology Adoption Lifecycle.

Technology Readiness Level — (TRL) A representation of an assessment of the maturity of an evolving technology. Different sectors and different countries use different TRL scales but the general philosophy is the same and they are each based on a linear model of innovation while at the same time recognizing that innovation is seldom linear, not all development cycles are the same, and the development of any particular technology may skip some readiness levels. The USA, Canada, Australia, and the European Commission, among others, use similar scales ranging from 1 to 9 (*see* Table 9 and Figure 5). Research and Technology Organizations work across the full TRL span but generally focus on developing technologies from about TRL 2 or 3 through to about TRL 7 or 8. Seven- and four-point scales are also in common use. Possibly the simplest TRL scale is TRL 1: Discovery, TRL 2: Development, TRL 3: Demonstration, TRL 4: Deployment. *See* References [146, 147, 148]. *See also* Proof of Concept, Speculative Concept, Technology Readiness, Technology Readiness Index, Technology Acceptance Model, Commercial Readiness Index, Technology Adoption Lifecycle, Technology-Translation Gap.

Table 9: A Generalized Description of Technology Readiness Levels (TRLs)

Technology Readiness Level	Description
TRL 1	Basic principles have been observed and reported and are becoming translated into applied research and development.
TRL 2	Practical applications and inventions are being identified.
TRL 3	Applied research and development are underway at laboratory scale, including proof of concept.
TRL 4	Multiple technological components, if applicable, are integrated and demonstrated to work together, again at laboratory scale.
TRL 5	The technological components are integrated for testing and validation in a simulated and/or realistic environment beyond the laboratory.
TRL 6	The integrated technological components in a model or prototype are tested and validated in a simulated and/or realistic environment beyond the laboratory.
TRL 7	A complete prototype, at or near full-scale, is ready for demonstration and/or demonstrated in a realistic operational environment.
TRL 8	A complete technology has been tested and demonstrated to work in its final form and under realistic operational conditions.
TRL 9	A complete technology, in its final form and including any final "fixes," has been proven through deployment in actual operational environments and conditions.

Technology S-curve — *See* S-curve.

Technology Stage-Gate™ Process — (TechSG Process) Like the traditional Stage-Gate® process but designed specifically for technology development programs, and having more emphasis on managing through the development steps lying within the fuzzy front-end (FFE) of the process. Reference [149]. *See also* Idea-to-Launch Process, Stage-Gate® Product Development Process.

Technology Transfer	*See* Technology Dissemination.
Technology-Translation Gap	The gap between a technology that is at an early stage of development (i.e., at the discovery, technology concept, or invention stage) and the product development stage (from which a prototype could be developed). In terms of Technology Readiness Levels, this would be the gap between a technology at TRL 2-3 and at TRL 5-6. *See also* Technology Readiness Level.
Technopole	*See* Innovation Ecosystem.
Technopolis Complex	*See* Innovation Ecosystem.
Technostructure	The management cadres of corporations that, taken together, are sometimes viewed as being the planning and decision-making agents of advanced capitalist economies. From this point of view, the technostructure strongly influences, guides, and/or determines economic development in an economy. Reference [32].
Technovation	A *portmanteau* or mixed word derived from the words technological and innovation. In the world of technological innovation, it is usually used to represent the process of technological innovation. Caution should be exercised, however, as it is also frequently used with other meanings, such as in educational and mentoring programs, for example.
TechSG Process	*See* Technology Stage-Gate™ Process.
TEEPSE Futures	Technological, economic, environmental, political, social, and/or ethical future possible events. *See* Foresight.
Ten Cs	The "*Ten Cs*" for successful implementation of technological innovations, as described by Rothwell, are "*effective Communications to gain Consensus for Change, Champions to sustain Continuous Commitment to Change, a Culture that is Customer Centred.*" Reference [102]. *See also* Principles of Innovation.
Tenets of Innovation	See Principles of Innovation.

Ten/Five Rule	*See* 10/5 Rule.
Term Sheet	A document that sets out terms and conditions of a proposed contractual agreement between two parties around the sale of a specific technology, line of business, or business. A term sheet can be binding or non-binding. The term sheet is used to guide the preparation of a final contractual agreement that is binding. *See also* Boilerplate.
TFP	*See* Total Factor Productivity.
The Borg Law	Essentially a strategy that a "*laggard*" company should, or must, leapfrog the industry leader by absorbing and then extending the leader's advantage, in order to compete effectively in a marketplace. Example: Microsoft's adoption of the features of *Netscape* and subsequent improvements to better position *Internet Explorer*. Reference [54]. *See also* Davidow's Law.
The Chasm	*See* Tipping Point.
Theory of Inventive Problem Solving	(TIPS, or TRIZ) A systematic approach to the process of invention developed by Genrich Altshuller in the 1960s, based on a comprehensive analysis of the patented solutions to hundreds of thousands of previously solved inventive problems. He identified a series of approaches that, taken together, were key to the solution of the majority of these problems. Altshuller called his approach the "*Theory of Inventive Problem Solving*" (TIPS, or TRIZ, the Russian acronym). The TRIZ approach involves several groups of methods that can be used to look at a problem in ways that, either individually or in combination, frequently lead to finding an inventive solution. *See also* Altshuller, Creative Thinking Models. Reference [17].
Theory of the Firm	Any of several economic theories that attempt to explain the nature of corporations, including their existence, behavior, structure, and their relationship to other corporations and the marketplace. Most such economic theories date back to the mid 1900s, although one of the first was Adam Smith's agency theory and his description of the efficiency of corporations (in his book *The Wealth of Nations*, 1776).

Thinking Out-of-the-Box — A synonym for Lateral Thinking. *See* Creative Thinking Models.

Think Up — A synonym for Brainstorming. *See* Creative Thinking Models.

Third Generation Innovation Model — The Coupling Model, *See* Generations of Innovation, Non-Linear Innovation Models.

Third-Generation Nanotechnology — Molecular nanotechnology. *See* Generations of Nanotechnology.

Third Generation of Technology Foresight — *See* Generations of Technology Foresight.

Third Industrial Revolution — *See* Technological Ages.

Third Mission — Participation by universities in working with governments and industry to enable economic growth (via innovation) and social progress has been referred to as "*Mode 2 Research*" and the "third mission" (after teaching and research) of an "*Entrepreneurial University.*" For universities that have transitioned to a "third-mission" strategy, reference is sometimes made to a shift from an "*endless frontier*" of discovery (Mode 1) research as an end unto itself to an "*endless transition*" in which discovery research is translated into applied (Mode 2) research and development, and thence into commercial deployment and use. Reference [150]. *See also* Entrepreneurial University, Modes of Science, Innovation Ecosystem Models, Triple-Helix Model, Quad-Helix Model.

Third Wave — *See* 3rd Wave.

Thought Leader — A person that demonstrates leadership in the communication of leading futuristic, strategic, and/or systems thinking to key stakeholders in a field of endeavour. Thought leaders are frequently advocates for and/or leaders of change in a field and are generally perceived to be "*innovative*" in the broadest sense of the term.

TH Theory — *See* Triple-Helix Model.

Timeboxing — A project or time-management technique in which a fixed time period ("time box") is set for the completion of a project, usually including the delivery of project deliverables. This tool is sometimes used in lean innovation and lean manufacturing.

Time to Market — (TTM) In general, the elapsed time between origination of the concept for a new product, process, or service, and the point at which it is available for sale in the marketplace. Depending on the organization and the sector(s) in which it operates the definition of the starting time can vary, such as when an idea is recorded, or when a product development project is approved or staffed. Ambiguity about the definitive start time has, on part, led to the term "*fuzzy front-end.*" Similarly, the definition of the ending time can vary, such as when a first unit is shipped or purchased, or when a specified production rate. TTM can be considered to be a product development process metric. Also termed Cycle Time, Rate of Innovation, Speed to Market. The term "*Fast Innovator*" refers to a person, team, or organization that is able to achieve reduced and/or short times to market. *See also* Fuzzy Front-End.

Tipping Point — In general, a "*tipping point*" refers to a transition point beyond which some kind of "*critical mass*" has been reached, and an idea, trend, social behavior, or product sales level spreads and/or increases very dramatically. Also referred to as a "*turning point.*" Example: In epidemiology, the tipping point is when an infectious disease spreads beyond the capacity of any localized efforts to bring it under control. In innovation, it is often described in terms of Rogers' diffusion of innovation model, as the transition between the stage of early adoptions (characterized by the innovators and the early adopters) and the next two large waves of adoptions (characterized by the early and late majorities). In this context, it is sometimes referred to as "*The Chasm.*" *See* Figure 6. *See also* J-curve.

TIPS — *See* Theory of Inventive Problem Solving.

T-KIBS — Technological KIBS. *See* Knowledge-Intensive Business Services.

TMI — Technology Market Intermediaries. *See* Knowledge-Intensive Business Services, Intermediary Organization.

Tool — *See* Technology.

Top-Down Economic Development
: An approach to economic development by which government tries to "pick winners," whether in terms of choosing particular applied R&D pathways, products, processes, or services, companies, or even clusters. This is typically done through the direct provision of grants or subsidies. This approach is generally viewed as being inefficient and/or ineffective. Conversely, industry-led approaches ("*Bottom-Up Economic Development*"), that are supported by government, are generally viewed as being more efficient and effective, especially in the case of Innovative Regional Clusters.

Top-Down Innovation
: The process of seeking-out, identifying, and evaluating potential market opportunities and then challenging the organization to come up with and develop concepts for innovative new products, processes, or services that could align with the selected market opportunities. *Bottom-Up Innovation*, on the other hand, refers to the process of originating and developing concepts for new business products, processes, or services and then evaluating them for market potential. *See also* Innovation, Upstream Innovation.

Total Factor Productivity
: (TFP) An economic measure of the contribution of technological innovation to labor productivity. In this case, labor productivity is divided into three principal components representing contributions from educational attainment, capital input, and total factor productivity (TFP), where TFP represents the influence of technological innovation. Also termed "Solow's Residual." Reference [151]. *See also* Multifactor Productivity.

Total Operating Expense
: *See* Capital Expense.

TPP Innovating Firm
: An OECD term for an organization that has implemented new or significantly improved technological products, processes, or services. The acronym "TPP" stands for "Technological Product and Process." Reference [8]. *See also* Innovation (20th–21st century).

TPP Innovation
: Technological Product and Process Innovation. A technological innovation involving a product or process. *See* Innovation (20th–21st century).

Trade Dress — The physical appearance of a product and/or its packaging. If the physical appearance of a product is unique, unusual, and/or widely recognized by the public, then its Trade Dress may be legally protected in the countries within which they are registered, and specific registration may not be needed. *See also* Intellectual Property.

Trade Secret — Any kind of technological or business know-how that is protected from competitors by keeping it confidential. *See also* Intellectual Property, Acquisition of Technology.

Trademark — A word, phrase, name, design, symbol, logo, or device associated with a business brand, product, or process, and which is used for differentiation from other brands or products in the same general line of business. The term *service mark* is sometimes used when a business service is meant. Registered trademarks and service marks are legally protected in the countries within which they are registered. *See also* Brand, Intellectual Property.

Trademarks — (Innovation Indicator) An indicator of the total number of new products and services being protected in an economy is the number of trademarks registered per 10,000 personnel in the labor force. *See* Innovation Indicators and Tables 4 and 8.

Transcendentalist Model — A model for the innovation process in which innovation arises from an instantaneous act or inspiration of genius. *See also* Cumulative Synthesis Model, Mechanistic Model.

Transformational Innovation — *See* Disruptive Innovation.

Transilience — A term coined by Abernathy and Clark [13] to describe the potential influence of a technological innovation on an organization's prior technological knowledge and resources, and its influence on the competitive marketplace. The potential influences are represented as quadrants in a diagram termed a *Transilience Map*, which denotes "regular (i.e., incremental), niche, "revolutionary" (i.e., evolutionary), and "architectural" (i.e., disruptive) innovations. *See* Figure 1.

Transilience Map — *See* Transilience.

Transnational Corporation — *See* Multinational Enterprise.

TRI — *See* Technology Readiness Index.

Triadic Patent Family — *See* Patent Family.

Triadic Patents Issued — (Innovation Indicator) An indicator of the number of practical inventions with international potential that are being protected is the number of patents in triadic patent families issued per 1 million people in the population of a given economy. Triadic patent families are groups of patents that have been granted in multiple (three or more) countries on the same invention. *See* Innovation Indicators and Tables 4 and 8.

Trialability — An older term referring to the extent and ease with which a technological innovation can be tested and/or experimented with, i.e., tried, by a prospective customer before committing to a purchase. The term appears to have originated with Rogers and his diffusion of innovation model. Reference [31].

Trial and Error Strategy — *See* Rationalist Strategy.

Trialing — *See* Awareness, Trial, Availability, Repeat Model.

Trickle-Up Innovation — *See* Reverse Innovation.

Triple-Helix Field Theory — *See* Triple-Helix Model.

Triple-Helix Model — The "*Triple-Helix Model*" is an innovation ecosystem (sociological) model developed by Etzkowitz and Leydesdorff in 1994 to describe the roles and intersections of governments, universities, and industry in advancing knowledge-based economies [152, 153, 154]. By extension, the model has also been applied to the advancement of economies based on innovation (the *innovation-related triple-helix*). Here governments provide a regulatory framework, science and technology (S&T) policies, programs, and infrastructure; universities provide new knowledge, understanding, and technology, and industry provides production and deployment of the innovations into the marketplace. In the triple-helix model, universities would be the primary incubators for technology-based start-ups and small- and medium-sized enterprises (SMEs). The helix is a metaphor for an innovation process that is more like a spiral than a straight line or S-curve, and along the spiral there are multiple points of intersection and interaction among the three kinds of organizations. Synonyms include Triple-Helix Field Theory (or TH Theory), and the ABG Model (referring to academia, business, and government). In a different extension from innovation, there is also a sustainability-related triple-helix in which the public-at-large is substituted for industry, to represent societal concerns about technologies, industry, the environment, and sustainability. *See* Figure 8. *See also* Innovation Ecosystem Models, Innovation Ecosystem, Quad-Helix Model, N-Tuple-Helix Model.

TRIZ — *See* Theory of Inventive Problem Solving.

TRL — *See* Technology Readiness Level.

TRL 1 — Technology Readiness Level 1. Different countries and industrial sectors use somewhat different definitions but on nine-point TRL scales, TRL 1 generally refers to the stage at which basic principles have been observed and reported, and are becoming translated into applied research and development. *See* Table 9, Figure 5, and Technology Readiness Level.

TRL 2 — Technology Readiness Level 2. Different countries and industrial sectors use somewhat different definitions but on nine-point TRL scales, TRL 2 generally refers to the stage at which practical applications and inventions are being identified. *See* Table 9, Figure 5, and Technology Readiness Level.

TRL 3 — Technology Readiness Level 3. Different countries and industrial sectors use somewhat different definitions but on nine-point TRL scales, TRL 3 generally refers to the stage at which applied research and development are underway at laboratory scale, including proof of concept. *See* Table 9, Figure 5, and Technology Readiness Level.

TRL 4 Technology Readiness Level 4. Different countries and industrial sectors use somewhat different definitions but on nine-point TRL scales, TRL 4 generally refers to the stage at which multiple technological components, if applicable, are integrated and demonstrated to work together, again at laboratory scale. *See* Table 9, Figure 5, and Technology Readiness Level.

TRL 5 Technology Readiness Level 5. Different countries and industrial sectors use somewhat different definitions but on nine-point TRL scales, TRL 5 generally refers to the stage at which the technological components are integrated for testing and validation in a simulated and/or realistic environment beyond the laboratory. *See* Table 9, Figure 5, and Technology Readiness Level.

TRL 6 Technology Readiness Level 6. Different countries and industrial sectors use somewhat different definitions but on nine-point TRL scales, TRL 6 generally refers to the stage at which the integrated technological components in a model or prototype are tested and validated in a simulated and/or realistic environment beyond the laboratory. *See* Table 9, Figure 5, and Technology Readiness Level.

TRL 7 Technology Readiness Level 7. Different countries and industrial sectors use somewhat different definitions but on nine-point TRL scales, TRL 7 generally refers to the stage at which a complete prototype, at or near full-scale, is ready for demonstration and/or demonstrated in a realistic operational environment. *See* Table 9, Figure 5, and Technology Readiness Level.

TRL 8 Technology Readiness Level 8. Different countries and industrial sectors use somewhat different definitions but on nine-point TRL scales, TRL 8 generally refers to the stage at which a complete technology has been tested and demonstrated to work in its final form and under realistic operational conditions. *See* Table 9, Figure 5, and Technology Readiness Level.

TRL 9 Technology Readiness Level 9. Different countries and industrial sectors use somewhat different definitions but on nine-point TRL scales, TRL 9 generally refers to the stage at which a complete technology, in its final form and including any final "fixes", has been proven through deployment in actual operational environments and conditions. *See* Table 9, Figure 5, and Technology Readiness Level.

TTM — *See* Time to Market.

Turning Point — *See* Tipping Point.

Type I Error — *See* Decision-Making Errors.

Type II Error — *See* Decision-Making Errors.

U

Ubiquitous Computing — *See* Internet of Things.

UILO — University-Industry Liaison Office. *See* Industry Liaison Office.

Unarticulated Needs — Products, processes, or services that customers might be willing to buy, but which they are unable or unwilling to imagine and/or describe.

Universal Success Curve — A graph illustrating the 1997 results of research by Stevens and Burley, which led them to conclude that in most industries, it takes about 3,000 initial, undeveloped ideas to produce a single successful commercial product. They also identified the improving success rates as an idea survives various intermediate staged-gates of development, as illustrated in Figure 19. Others have found similar trends but somewhat different numbers for different industries, such as an estimate that in the pharmaceutical industry, it could take at least 6,000 ideas to produce a commercially successful product. References [155, 156]. *See also* Rainmaker.

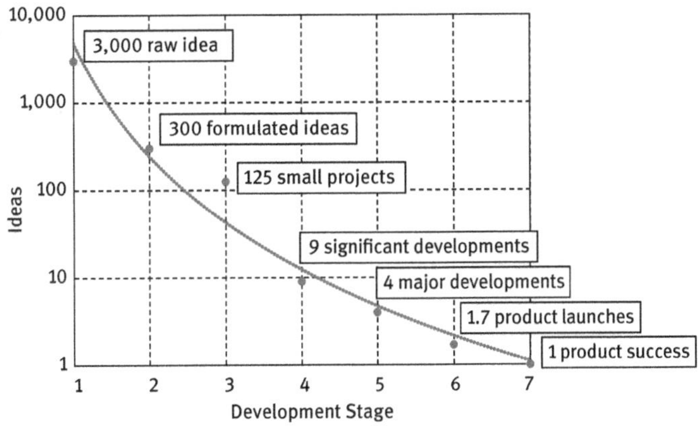

Figure 19: Illustration of a "*Universal Success Curve.*" From data reported in Stevens and Burley, 1997 [155]. The dotted line represents a power law fit of the reported data.

DOI 10.1515/9783110429176-022

University-Industry Liaison Office	(UILO) *See* Industry Liaison Office.
Upstream Innovation	The process of originating, evaluating, and developing concepts for new business products, processes, or services. This can include the process of discovery. *Downstream Innovation*, on the other hand, refers to the process of converting such concepts into market-ready products, processes, or services and then introducing them into the marketplace. This can include the process of invention. *See also* Innovation, Top-Down Innovation.
Upstream Research	*See* Research and Development.
Upstream Studies	*See* Research and Development.
User Innovation	*See* Consumer-Innovators.
Utility Patent	*See* Patent.

Valley of Death (Innovation) As technology development proceeds, there is a tendency for financing, usually government financing, to become increasingly difficult to obtain, as the technology is developed from basic and applied research through to speculative concept, engineering development, and proof of concept (i.e., from Technology Readiness Levels TRL 1 through 3). Other kinds of financing such as venture capital, private equity, debt financing tend to become available as the technology matures from field demonstration and first deployment through to full maturity in competitive markets (i.e., from Technology Readiness Level TRL 9 through Commercial Readiness Index level CRI 6). In between lies the so-called *Valley of Death*, a series of technology development steps such as prototype development and testing, scale-up, and pilot testing (i.e., from Technology Readiness Levels TRL 4 through 8) for which financing is the most difficult to obtain. See Figure 20.

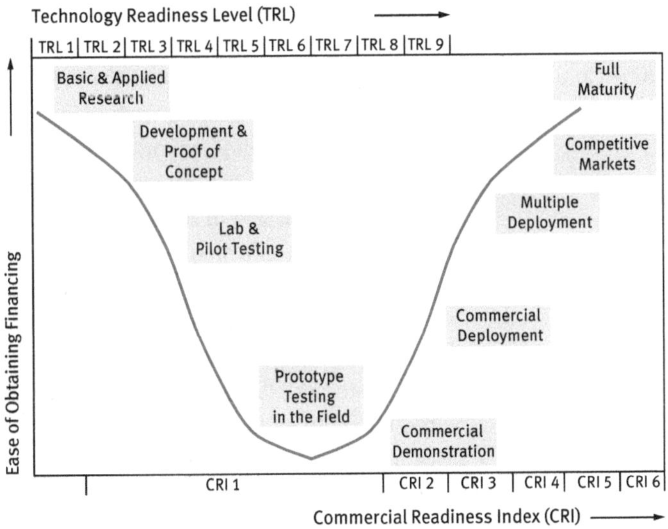

Figure 20: Illustration of Technology Development Stages and the Valley of Death.

DOI 10.1515/9783110429176-023

(Technological vs. Commercialization) Within the *Valley of Death*, two sub-categories are sometimes distinguished. The *Technological Valley of Death* refers to early-stage technology development phases such as speculative concept development, engineering development, and proof of concept. The *Commercialization Valley of Death* refers to later-stage technology development phases such as full field- or plant-scale commercial demonstrations and first and second customer sales and deployments. The valleys of death are a challenge for innovative organizations of all kinds, but they particularly plague research and technology organizations (RTOs) and start-up companies. See Figure 20.

(Entrepreneurship) In entrepreneurship, the Valley of Death refers to the gap, or period of time, between the initial funding and launch of a start-up company and the point at which it starts generating revenue from sales. During this interval, start-up companies tend to have a high risk of requiring more operating capital than they can afford, which can cause the enterprise to fail. An illustration showing steps in a linearized process leading from the initial business concept through to a successful, sustainable company is sometimes termed a Death Valley Curve. Also termed Death Valley.

Value Chain

A systems view of the series of processes by which a manufacturer begins with starting materials, adds value, and profitably sells a (higher value) final product, process, or service. Each such process comprises a subsystem involving subsystems, each with its own inputs, use of resources, transformation processes, and outputs. In Porter's model, value chains have five main segments: "*Inbound Logistics*" (receiving and preparing the raw materials), "*Operations*" (converting the raw materials into a finished product, process, or service), "*Outbound Logistics*" (delivering it to a customer), "*Marketing and Sales*," and "*Service*." The name has been attributed to Michael Porter, who developed a specific value chain model. Reference [117]. Within a value chain, the processes occurring with the company are sometimes referred-to as "*Business Logistics Management*," while the process handled by external firms are sometimes referred-to as "*Supply Chain Management*." Such supply chain firms range from raw material suppliers, to assemblers, to shippers, to retailers.

Venture Capital	Final-stage, usually substantial, capital for a commercialization process that is typically raised through formal, institutional investment firms. Venture capital is usually used to take a product, process, or service that has already been introduced into the marketplace and to rapidly drive it "up" past the marginal or break-even level and into profitability. *See* 10/5 rule. *See also* Capital, Sweat Equity, Seed Capital, Start-Up Capital, Pre-Venture Capital.
Venture Capital Intensity	The ratio of venture capital investments to gross domestic product in an economy. *See* Innovation Indicators and Table 4.
Venturing	Creating a for-profit business enterprise of some kind. A commercialization strategy that involves venturing involves creating or expanding a business in order to realize an innovation.
Verhulst Curve	*See* S-Curve (Science).
Verhulst, Pierre François (1804–1849)	A Belgian mathematician known to the innovation world for his development of the sigmoid S-curve, which he originally used to describe population growth. The concept of a growth rate that is initially exponential but then slows and plateaus has since been applied to many other phenomena, including product lifecycles. The S-curve is sometimes termed a Verhulst Curve. *See* S-curve.
Vertical Thinking	A synonym for *Linear Thinking*. *See* Creative Thinking Models.
Virtual Cluster	*See* Business Ecosystem.
Visionary	*See* Early Adopter.
VoC	*See* Voice of the Customer.
Voice of the Customer	(VoC) Feedback of almost any kind received from customers of an existing or proposed product, process, or service. Such feedback can be integrated into any stage of the product development process, from the conception of prospective new products to the marketing and sales of existing ones.
Voucher Program	*See* Innovation Voucher Program.

W

Water-Ribbed Balloon	A metaphor for a model of the product development process that is very similar to the stage-gate process for managing technological innovation. Reference [157].
Waterfall Method	A new product development process in which a linear series of developmental steps is followed in sequence and with each step only being undertaken after the preceding one has been completed. This method seems to have originated in the software industry, but it has been used more broadly in the manufacturing and other industries as well. Modern usage of this method often included feedback loops that permit making backing up into earlier steps to make changes as issues are discovered downstream. *See also* New Product Development Process, Stage-Gate® Product Development Process. References [4,5].
Waves of Innovation	*See* Kondratieff Waves.
Weak Signals	*See* Foresight.
WEF	World Economic Forum (Geneva, Switzerland). *See* Global Competitiveness Index, Competitiveness Drivers.
White Space	*See* White Space Mapping.
White Space Mapping	(Business Processes) In business processes, *White Space* usually refers to areas within which policies and/or authorities are undefined or vague. These create inefficiencies and mapping the *White Space* can help identify opportunities for improvement and/or creativity. The term *White Space* seems to have originated with the concept of the white space between the boxes in an organizational chart. Areas within which policies and/or authorities are defined and/or clear are sometimes referred to as *Black Space*. (Business Development) In business development, *White Space* usually refers to either new product/process/service

DOI 10.1515/9783110429176-024

offerings to current customers or to new or existing product/process/service offerings to new customers. Either way, *White Space* refers to the aspect that is new. In some usage, the term is used to refer to market niches that are under-served, but this is probably better referred to as *Grey Space*. In this context, *White space* could also refer to a customer/market opportunity for which an appropriate product/process/service is not available.

Whole Product R&D — The practice of keeping research and development (R&D) involved through all stages of a product's launch, evolution, marketing, and lifecycle in order to ensure that marketing and customer feedback is used to continuously improve, if possible, the product, process, or service in order to extend its lifecycle. The value in keeping the R&D function involved is that it can be difficult for others to discern the difference between a technologically simple change and a substantial technological boundary. Reference [160]. *See also* Research and Development.

Wild Cards — *See* Foresight.

Wisdom Hierarchy — An idealized linear model for the process by which information is obtained from data, knowledge from information, and wisdom from knowledge. *Also termed* Data-Information-Knowledge-Wisdom Hierarchy, DIKW Hierarchy, Data Hierarchy, Information Hierarchy, Knowledge Hierarchy.

Wizard — *See* Rainmaker.

Working Model — A model built to demonstrate that a product or process has been reduced to practice and basically works (proof of concept). Working models are expected to function, but are usually not built to scale and their functioning has usually not been optimized. *See also* Prototype. Reference [6].

"Works Like" Model — A model that illustrates the function of a product, process, or service without necessarily working exactly like the final product will. As such, a "works like" model is more advanced than a mockup, but less representative of the final product than a working model, engineering prototype, or production prototype. The benefits of a "works like" model are the ability to quickly and inexpensively produce a model that can be shown to prospective customers in order to get feedback before finalizing the product design. Similarly, a "looks like" model illustrates the appearance of a product, process, or service without functioning exactly like the final product will. *See also* Prototype, Fail Fast.

World-First

In discussing the introduction to the marketplace of new products, processes, or services, it can sometimes be useful to categorize them as "*Firm-First,*" "*Country-First,*" or "*World-First.*" Some companies' competitive markets are such that Country-First is sufficient for their competitive needs, whereas companies that compete globally tend to aim for World-First.

Y

Year 1 New Product Sales Index	*See* New Product Vitality Index.
Yield	*See* Return on Investment.

DOI 10.1515/9783110429176-025

Z

Zizhu Chuangxin Self-Determined Innovation. *See* Indigenous Innovation.

DOI 10.1515/9783110429176-026

References

[1] Godin, B., *Innovation Contested. The Idea of Innovation Over the Centuries*, Routlege: New York, 2015.

[2] Godin, B., "The Vocabulary of Innovation: A Lexicon," Presented at 2nd CASTI Workshop, Agder, Norway, October 20, 2014.

[3] Schumpeter, J.A., *Business Cycles. A Theoretical, Historical and Statistical Analysis of the Capitalist Process*, McGraw-Hill: New York, 1939.

[4] Schumpeter, J.A., *Capitalism, Socialism and Democracy*, 3rd Ed., HarperCollins: New York, 1950.

[5] Touhill, C.J.; Touhill, G.J.; O'Riordan, T.A., Commercialization of Innovative Technologies," Wiley-Interscience: New York, 2008.

[6] Lux, D.; Rorke, M., *From Invention to Innovation*, Report DOE/GO-10099-810, US Department of Energy: Washington, DC, 1999.

[7] OECD, *Frascati Manual, Proposed Standard Practice for Surveys on Research and Experimental Development*, Organisation for Economic Co-Operation and Development (OECD): Paris, 2002.

[8] OECD, *Oslo Manual, Guidelines for Collecting and Interpreting Innovation Data*, 3rd Ed., Organisation for Economic Co-Operation and Development (OECD): Paris, 2005.

[9] Glasco, J., *The Evolving Language of Innovation, A Glossary of Terms*, GCA Thought Leader Publications: USA, 2013.

[10] Thota, H.; Munir, Z., *Palgrave Key Concepts: Key Concepts in Innovation*, Macmillan: New York, 2011.

[11] Wilenius, M.; Kurki, S., *Surfing the Sixth Wave. Exploring the Next 40 Years of Global Change*, Finland Futures Research Centre, University of Turku: Helsinki, December 2012, 127 pp.

[12] 3M, *3M Innovation Story: Uncommon Connections, Innovative Solutions*, 3M Inc.: Minneapolis, 2011.

[13] Abernathy, W.J.; Clark, K.B., "Innovation: Mapping the Winds of Creative Destruction," *Res. Policy*, **1985**, *14(1)*, 3–22.

[14] Christensen, C.M., "*The Innovator's Dilemma: When New Technologies Cause Great Firms to Fail*," Harvard Business Review Press: Boston, 1997.

[15] Freller, L. "Alpha vs. Beta Testing," Blog Post, Centercode, Laguna Hills, CA, 17 January 2011, http://www.centercode.com/blog/2011/01/alpha-vs-beta-testing/.

[16] 360logicaadmin, "What are Alpha, Beta and Gamma Testing?" The Official 360logica Blog, 27 June 2012, http://www.360logica.com/blog/2012/06/what-are-alpha-beta-and-gamma-testing.html.

[17] Altshuller, G., *And Suddenly the Inventor Appeared*, English translation, Technical Innovation Center: Worcester, MA, 2004.

[18] Ansoff, I., "Strategies for Diversification," *Harv. Bus. Rev.*, **1957**, *35(5)*, 113–124.

[19] Jaruzelski, B.; Dehoff, K., "How the Top Innovators Keep Winning," *Strategy+Business*, **2010**, *61 (Winter)*, 16 pp.

[20] Henderson, R.M.; Clark, K.B., "Architectural Innovation: The Reconfiguration of Existing Product Technologies and the Failure of Established Firms," *Admin. Sci. Q.*, **1990**, *35(1)*, 9–30.

[21] Thota, H.; Munir, Z., *Palgrave Key Concepts: Key Concepts in Innovation*, Macmillan, NY, 2011.

[22] Leydesdorff, L.; Meyer, M., "The Decline of University Patenting and the End of the Bayh-Dole Effect," *Scientometrics*, **2010**, *83(2)*, 355–362.

[23] Bush, V., "Science, the Endless Frontier. A Report to the President on a Program for Postwar Scientific Research," July, 1945; reprinted July 1960 by the National Science Foundation, Washington. https://ia600207.us.archive.org/12/items/scienceendlessfr00unit/scienceendlessfr00unit.pdf.

[24] Reingold, N., *American National Biography*, Oxford University Press: Oxford, 2010.

DOI 10.1515/9783110429176-027

[25] Dedehayir, O.; Seppänen, M., "Birth and Expansion of Innovation Ecosystems: A Case Study of Copper Production," *J. Technol. Manag. Innov.*, **2015**, *10(2)*, 145–153.
[26] Foster, R.N., *Innovation: The Attacker's Advantage*, Summit Books: New York, 1986.
[27] Kline, S.J., "Innovation is Not a Linear Process," *Res. Manage.*, **1985**, *Jul.–Aug.*, 36–45.
[28] Kline, S. J.; Rosenberg, N., "An Overview of Innovation," in Laudan, R.; Rosenberg, N. (Eds.), *The Positive Sum Strategy: Harnessing Technology for Economic Growth*, National Academies Press: Washington, 1986, pp. 275–305.
[29] Chesbrough, H.W., *Open Innovation: The New Imperative for Creating and Profiting from Technology*, Harvard Business School Press: Boston, 2003.
[30] Huizingh, E.K.R.E., "Open Innovation: State of the Art and Future Perspectives," *Technovation*, **2011**, *31(1)*, 2–9.
[31] Rogers, E.M., *Diffusion of Innovations*, 5th Ed., Free Press: New York, 2003.
[32] Courvisanos, J., "Technological Innovation: Galbraith, the Post Keynesians, and a Heterodox Future," *J. Post Keynesian Econ.*, **2005**, *28(1)*, 85–104.
[33] Mercier-Laurent, E., *Innovation Ecosystems,* Wiley: NY, 2011.
[34] ARENA, *Commercial Readiness Index for Renewable Energy Sectors*, Australian Renewable Energy Agency: Canberra, February, 2014.
[35] de Castro, G.M.; Verde, M.D.; Sáez, P.L.; López, J.E.N., *Technological Innovation, An Intellectual Capital-Based View*, Palgrave Macmillan: London, 2010.
[36] "The Global Competitiveness Report," World Economic Forum, Geneva, Switzerland, 2014.
[37] Von Hippel, E.; Ogawa, S.; De Jong, J.P.J., "The Age of the Consumer-Innovator," *MIT Sloan Mgmt. Rev.*, **2011**, *53(1)*, 27–35.
[38] Cooper, R.G., *Winning at New Products: Creating Value Through Innovation*, 4th Ed., Basic Books: New York, 2011.
[39] Cooper, R.G., "New Products – What Separates the Winners From the Losers and What Drives Success," *The PDMA Handbook of New Product Development*, Kahn, K.B. (Ed.), 3rd Ed., Wiley: New York, 2013, pp. 3–34.
[40] Osborn, A.F., *Your Creative Power, How to Use Imagination*, Charles Scribner's Sons: New York, 1949.
[41] Osborn, A.F., *Applied Imagination: Principles and Procedures of Creative Problem-Solving*, Charles Scribner's Sons: New York, 1953.
[42] Isaksen, S., Treffinger, D., "Celebrating 50 Years of Reflective Practice: Versions of Creative Problem Solving," *J. Creative Beh.*, **2004**, *38(2)*, 1–27.
[43] Parnes, S.J., *Creative Behavior Guidebook*, Scribner: New York, 1967.
[44] DeBono, E., *The Use of Lateral Thinking*, Jonathan Cape: London, 1967 (also published as *New Think*, Basic Books: New York).
[45] DeBono, E., *Lateral Thinking for Management,* American Management Association: New York, 1972.
[46] Gordon, W.J.J., *Synectics, The Development of Creative Capacity*, Harper and Row: New York, 1961.
[47] Nolan, V., "Whatever Happened to Synectics?" *Creativity Innov. Mgmt.*, **2003**, *12(1)*, 24–27.
[48] Ledford, H. "Riding the CRISPR Wave. Biologists are Embracing the Power of Gene-Editing Tools to Explore Genomes." *Nature*, **2016**, *531*, 156–159.
[49] Talbot, D., "Precise Gene Editing in Plants," *MIT Technol. Rev.*, **2016**, *119(2)*, 41.
[50] Lyons, J.; Chait, R.; Long, D., "Critical Technology Events in the Development of Selected Army Weapons Systems: A Summary of 'Project Hindsight Revisited,' Defense & Technology Paper DTP-035, National Defense University Center for Technology and National Security Policy: Washington, DC, Sept. 2006.
[51] Usher, A.P., *A History of Mechanical Inventions*, 2nd Ed., Harvard Univ. Press: Cambridge, MA, 1954.
[52] Usher, A.P., "Technical Change and Capital Formation," In *Capital Formation and Economic Growth. A Conference of the Universities-National Bureau Committee for Economic Research,* National Bureau of Economic Research (Ed.), Princeton University Press: Princeton, NJ, 1955, pp. 521–548.

[53] Donaldson, K.M.; Ishii, K.; Sheppard, S.D., "Customer Value Chain Analysis," *Res. Eng. Des.*, **2006**, *16*, 174–183.
[54] Lewis, T., "Corporate R&D in the Age of E-Commerce," *Res. Technol. Mgmt.*, **2000**, *43(6)*, 16–20.
[55] Verganti, R., *Design-Driven Innovation. Changing the Rules of Competition by Radically Innovating What Things Mean,* Harvard Business Press: Boston, 2009.
[56] Huang, G.Q.; Mak, K.L., "Developing a Generic Design for X Shell," *J. Eng. Des.*, **1997**, *8(3)*, 251–260.
[57] Huang, G.Q.; Mak, K.L., "The DFX Shell: A Generic Framework for Applying 'Design for X' (DFX) Tools," Int. J. Computer Integrated Manuf., **1998**, *11(6)*, 475–484.
[58] Milford, R., "Design for Future Reuse," Low Carbon Materials Processing, Working Paper W2, Cambridge University: Cambridge, September, 2010, http://www.lcmp.eng.cam.ac.uk/wp-content/uploads/W2-Design-for-future-reuse.pdf.
[59] Furr, N.; O'Keeffe, K.; Dyer, J.H., "Managing Multiparty Innovation," *Harvard Bus. Rev.*, **2016**, *Nov.*, 76–83.
[60] Romer, P.M., "Increasing Returns and Long-Run Growth," *J. Political Econ.*, **1986**, *94(5)*, 1002–1037.
[61] Lucas, R.E., "On the Mechanics of Economic Development," *J. Monetary Econ.*, **1988**, 22(1), 3–42.
[62] Romer, P.M., "Endogenous Technological Change," *J. Political Econ.*, **1990**, 98(5), S71–S102.
[63] Etzkowitz, H., "Innovation in Innovation: The Triple Helix of University-Industry-Government Relations," *Soc. Sci. Inform.*, **2003**, *42(3)*, 293–337.
[64] OECD, "*A Guiding Framework for Entrepreneurial Universities*," Organisation For Economic Co-Operation and Development (OECD): Paris, 2012.
[65] Huenteler, J.; Schmidt, T.S.; Ossenbrink, J.; Hoffmann, V.H., "Technology Life-Cycles in the Energy Sector — Technological Characteristics and the Role of Deployment for Innovation," *Technol. Forecast. Soc. Change,* **2016**, *104*, 102–121.
[66] Courvisanos, J., "Technological Innovation: Galbraith, the Post Keynesians, and a Heterodox Future," *J. Post Keynesian Econ.*, **2005**, *28(1)*, 85–104.
[67] Popper, R., "Critical Factors Influencing the Selection of Foresight Methods," *myForesight*, **2011**, *(April)*, 13–26.
[68] Dalziel, M., "4th Pillar Organizations in Canada," Report for Prime Minister's Advisory Council on Science and Technology, Government of Canada, Sept. 15, 2005, http://papers.ssrn.com/sol3/papers.cfm?abstract_id=1518863.
[69] Andersen, A.D.; Andersen, P.D., "Innovation System Foresight," *Technol. Forecasting Social Change*, **2014**, *88*, 276–286.
[70] Chan, C.; Daim, T., "Exploring the Impact of Technology Foresight Studies on Innovation: Case of BRIC Countries," *Futures*, **2012**, *44*, 618–630.
[71] Ginarte, J.C.; Park, W.G. "Determinants of Patent Rights: A Cross National Study," *Research Policy*, **1997**, *26*, 283–301.
[72] Park, W.G. "International patent protection: 1960–2005," *Research Policy*, **2008**, *37*, 761–766.
[73] UAE, "Government Innovation Labs," Mohammed Bin Rashid Centre for Government Innovation, Dubai, United Arab Emirates, 2016, https://www.mbrcgi.gov.ae/lab.aspx.
[74] Storper, M., *The Regional World – Territorial Development in a Global Economy*, Guilford Press: New York, 1997.
[75] OECD, "ISIC Rev. 3 Technology Intensity Definition. Classification of Manufacturing Industries into Categories Based on R&D Intensities," OECD Directorate for Science, Technology and Industry, Organisation for Economic Co-operation and Development (OECD): Paris, France, July 7, 2011.
[76] Holland, M., *Industrial Explorers*, Harper & Bros.: New York, 1928.
[77] Godin, B., "The Linear Model of Innovation (II): Maurice Holland and the Research Cycle," Working Paper No. 3, Project on the Intellectual History of Innovation, Institut National de la Recherche Scientifique (INRS): Montreal, 20 pp., 2009.

[78] Storper, M., *The Regional World – Territorial Development in a Global Economy*, Guilford Press: New York, 1997.
[79] Godin, B.; Lane, J.P., "'Pushes and Pulls': The Hi(story) of the Demand Pull Model of Innovation," Working Paper No. 13, Project on the Intellectual History of Innovation, Institut National de la Recherche Scientifique (INRS): Montreal, 39 pp., 2013.
[80] Usher, A.P., *A History of Mechanical Inventions*, 1st Ed., McGraw-Hill: New York, 1929.
[81] Curley, M., "Twelve Principles for Innovation 2.0," *Nature*, **2016**, *533*, 314–316.
[82] AMA, *The Quest for Innovation. A Global Study of Innovation Management 2006–2016*, American Management Association: New York, 2006.
[83] Khan, Z.; Joseph, K., "Embracing the Paradoxes of Innovation," *Stanford Social Innov. Rev.*, **2013**, *Summer*, 21–23.
[84] OECD, "National Innovation Systems: Report on Pilot Case Studies," Organisation for Economic Co-operation and Development (OECD): Paris, France, 1996.
[85] Conference Board, "Performance and Potential 2003–04: Defining the Canadian Advantage," Special Report, The Conference Board of Canada: Ottawa, 2003.
[86] Conference Board, "Including Innovation in Regulatory Frameworks. 4th Annual Innovation Report," Report 361–02, Conference Board of Canada, Ottawa, 2002.
[87] Etzkowitz, H., "Innovation in Innovation: The Triple Helix of University-Industry-Government Relations," *Soc. Sci. Information*, **2003**, *42(3)*, 293–337.
[88] Rabe, C.B., *The Innovation Killer*, American Management Association: New York, 2006.
[89] Farson, R.; Keyes, R., *The Innovation Paradox: The Success of Failure, the Failure of Success*, Free Press: New York, 2003.
[90] Davila, T.; Epstein, M.J., *The Innovation Paradox: Why Good Businesses Kill Breakthroughs and How They Can Change*, Berrett-Koehler Publishers: San Francisco, CA, 2014.
[91] Conference Board, "'Don't Overlook the "MEs' Medium-sized Firms Show the Way to Global Success. 6th Annual Innovation Report," Report 670-04, Conference Board of Canada: Ottawa, 2005.
[92] Rapp, J.D., "Inside Whirlpool's Innovation Machine," Management Innovation eXchange Blog Post, 23 January, 2013, http://www.managementexchange.com/story/inside-whirlpools-innovation-machine.
[93] Tennant Snyder, N.; Duarte, D.L., *Unleashing Innovation: How Whirlpool Transformed an Industry*, Jossey-Bass: San Francisco, CA, 2008.
[94] Davis, P., "Innovation White Paper," January 2008, The Frost Research Center, Hope College, Holland, MI, downloaded 3 August 2016, http://www.hope.edu/frostcenter/Assets/InnovationWhitePaper.pdf.
[95] OECD, "*Innovation Vouchers*," OECD Innovation Policy Platform, Organisation For Economic Co-Operation and Development (OECD): Paris, February 2010.
[96] Howells, J., "Intermediation and the role of intermediaries in innovation," *Res. Policy*, **2006**, *35*, 715–728.
[97] Howard Partners Pty., "Study of the Role of Intermediaries in Support of Innovation," Report for Department of Industry, Tourism and Resources, April, 2007.
[98] Schmookler, J., *Invention and Economic Growth*, Harvard University Press: Cambridge, 1966.
[99] European Commission, "KETs: Time to Act," Final Report, High-Level Expert Group on Key Enabling Technologies, European Commission: Brussels, June, 2015.
[100] UNESCO, "Toward Knowledge Societies," UNESCO World Report, United Nations Educational, Scientific and Cultural Organization: Paris, 2005.
[101] Kondratieff, N., *Long Wave Cycle*, Translation of the original Russian book of 1925, E.P. Dutton: New York, 1984.
[102] Rothwell, R., "Successful Industrial Innovation: Critical Factors for the 1990s," *R&D Mgmt.*, **1992**, *22(3)*, 221–239.

[103] Sehested, C.; Sonnenberg, H., *Lean Innovation. A Fast Path from Knowledge to Value*, Springer-Verlag: Berlin, 2011.

[104] Browne, C.A., "Liebig and the Law of the Minimum," In *Liebig and After Liebig: A Century of Progress in Agricultural Chemistry*, Moulton, F.R. (Ed.), Am. Assoc. Adv. Sci.: Washington, DC, 1942, pp. 71–82.

[105] Godin, B., "In the Shadow of Schumpeter: W. Rupert Maclaurin and the Study of Technological Innovation," *Minerva*, **2008**, *46(3)*, 343–360.

[106] OECD, "Main Science and Technology Indicators," Organisation for Economic Co-Operation and Development (OECD): Paris, France, 2014.

[107] Schramm, L.L.; Nyirfa, W.; Grismer, K.; Kramers, J., "Research and Development Impact Assessment for Innovation-Enabling Organizations," *Can. Public Admin.*, **2011**, *54(4)*, 567–581.

[108] Johansson, F., *The Medici Effect. Breakthrough Insights at the Intersection of Ideas, Concepts and Cultures*, Harvard Business School Press: Boston, 2004.

[109] Council of Canadian Academies, "Innovation and Business Strategy: Why Canada Falls Short," Expert Panel on Business Innovation Report, The Council of Canadian Academies: Ottawa, ON, 2009.

[110] Merton, R.K., "Resistance to the Systematic Study of Multiple Discoveries in Science," *Eur. J. Sociology*, **1963**, *4(2)*, 237–282.

[111] Epstein, R.C., "Industrial Invention: Heroic, or Systematic?" *Q. J. Econ.*, **1926**, *40(2)*, 232–272.

[112] Badulin, N.A., "Innovation Snail, Triple Helix, and Other Circular Processes in Economics," In *Triple Helix Conference 2013*, 8–10 July 2013, 16pp., Triple Helix Assoc.: Torino, Italy, http://www.triplehelixconference.org/th/11/bic/docs/Papers/Badulin.pdf.

[113] Leydesdorff, L., "The Triple Helix, Quadruple Helix, ..., and an N-tuple of Helices: Explanatory Models for Analyzing the Knowledge-based Economy," *J. Knowledge Econ.*, **2012**, *3(1)*, 25–35.

[114] Leydesdorff, L., "N-Tuple of Helices," in *Encyclopedia of Creativity, Innovation, and Entrepreneurship*, Carayannis, E.G. (Ed.), New York: Springer, 2013, pp. 1400–1402.

[115] Chesbrough, H.; Van Alstyne, M., "Permissionless Innovation, *Comm. ACM* (Association for Computing Machinery), **2015**, *58(8)*, 24–26.

[116] Pontin, J., "The Rules of Innovation," *Technol. Rev.*, **2005**, *108(5)*, 12.

[117] Porter, M., *Competitive Advantage: Creating and Sustaining Superior Performance*, The Free Press, NY, 1985.

[118] McGrath, M.E., *Setting the PACE in Product Development*, 2nd Ed., Taylor & Francis: Milton Park, UK, 2011.

[119] Asplund, C., "Beyond 'Triple Helix' – Towards 'Quad Helix'," Internet blog post, 22 March 2012, Bearing Consulting, http://blog.bearing-consulting.com/2012/03/22/beyond-triple-helix-towards-quad-helix/.

[120] Wilson, E.J., "Forms and Dynamics of Leadership for a Knowledge Society: The Quad," Center for International Development and Conflict Management, University of Maryland: College Park, 2003, http://www.cidcm.umd.edu/leadership/quad2.pdf.

[121] Cooke, P., "Introduction," In *Regional Innovation Systems. The Role of Governances in a Globalized World*, Braczyk, H-J.; Cooke, P.; Heidenreich, M. (Eds.), UCL Press: London, UK, 1998, pp. 2–25.

[122] WAITRO, World Association of Industrial and Technological Research Organizations, http://www.waitro.org/.

[123] EARTO, European Association of Research and Technology Organizations, http://www.earto.eu/.

[124] I-CAN, Innoventures Canada Inc., http://www.i-can.ca/.

[125] Immelt, J.R.; Govindarajan, V.; Trimble, C., "How GE is Disrupting Itself," *Harv. Bus. Rev.*, **2009**, *87(10)*, 56–65.

[126] Govindarajan, V., Trimble, C., *Reverse Innovation: Create Far from Home, Win Everywhere*, Harvard Business Review Press, 2012.

[127] von Zedtwitz, M.; Corsi, S.; Søberg, P.V.; Frega, R., "A Typology of Reverse Innovation," *J. Prod. Innov. Manag.*, **2015**, *32(1)*, 12–28.
[128] Prahalad, C.K., "The Innovation Sandbox," *strategy+business*, **2006**, *44*, 62–73.
[129] Scherer, F.M., "Demand-Pull and Technological Innovation: Schmookler Revisited," *J. Indus. Econ.*, **1982**, *30(3)*, 225–237.
[130] Carter, C. F.; Williams, B. R., *Industry and Technical Progress: Factors Governing the Speed of Application of Science*, Oxford University Press: London, 1957.
[131] Foster, R.N., *Innovation: The Attacker's Advantage*, Summit Books: New York, 1986.
[132] Asthana, P., "Jumping the Technology S-Curve," *IEEE Spectrum*, **1995**, June, 49–54.
[133] Kucharavy, D.; De Guio, R., "Application of S-Shaped Curves," *Procedia Eng.*, **2011**, *9*, 559–572.
[134] Vinge, V. "The Coming Technological Singularity: How to Survive in the Post-Human Era," In *Vision 21. Interdisciplinary Science and Engineering in the Era of Cyberspace*, 1993 Symposium Proceedings, NASA Conf. Publ. No. 10129, NASA Lewis Research Center, March, 1993.
[135] Schramm, L.L., *Nano- and Microtechnology from A – Z: From Nanosystems to Colloids and Interfaces*, Wiley-VCH: Weinheim, Germany, 2014.
[136] Nobel Media, "Robert M. Solow – Biographical," Nobel Media, Stockholm, 2014, Accessed 13 July 2015 at http://www.nobelprize.org/nobel_prizes/economic-sciences/laureates/1987/solow-bio.html.
[137] The Economist, "The Growth of Growth Theory," *The Economist, 18 May*, **2006**.
[138] Acemoglu, D., *Introduction to Modern Economic Growth*, Chapter 2, Princeton University Press: Princeton, NJ, 2009, pp. 26–76.
[139] Chandy, R.; Prabhu, J., "Innovation Typologies," In *Wiley International Encyclopedia of Marketing*, Online Edition, Part 5, Wiley: New York, 2010.
[140] Johnston, R.E.; Bate, J.D., *The Power of Strategy Innovation,* American Management Association: New York, 2003.
[141] UNESCO, "Towards Knowledge Societies," UNESCO World Report, United Nations Educational, Scientific and Cultural Organization: Paris, 2005.
[142] Florida, R., *The Rise of the Creative Class*, Basic Books: New York, 2002. See also *The Rise of the Creative Class – Revisited*, Basic Books: New York, 2012.
[143] Maclaurin, W.R., "Innovation and Capital Formation in Some American Industries," in *Capital Formation and Economic Growth*, Universities-National Bureau, Princeton University Press: Princeton, NJ, 1955, pp. 549–578.
[144] Parasuraman, A., "Technology Readiness Index (TRI). A Multiple-Item Scale to Measure Readiness to Embrace New Technologies," *J. Service Res.*, **2000**, *2(4)*, 307–320.
[145] Parasuraman, A.; Colby, C.L., "An Updated and Streamlined Technology Readiness Index. TRI 2.0," *J. Service Res.*, published online before print June 22, 2014, doi: 10.1177/1094670514539730.
[146] PWGS Canada, "Technology Readiness Levels," Public Works and Government Services Canada: Ottawa, 2013, https://buyandsell.gc.ca/initiatives-and-programs/build-in-canada-innovation-program-bcip/program-specifics/technology-readiness-levels.
[147] EARTO, "The TRL Scale as a Research & Innovation Policy Tool, EARTO Recommendations," European Association of Research and Technology Organizations: Brussels, 2014.
[148] ARENA, "Technology Readiness Levels for Renewable Energy Sectors," Australian Renewable Energy Agency: Canberra, February, 2014.
[149] Ajamian, G.M.; Koen, P.A., "Technology Stage-Gate™: A Structured Process for Managing High-Risk New Technology Projects," in Belliveau, P.; Griffin, A.; Sommermeyer, S. (Eds.), *The PDMA ToolBook for New Product Development*, Wiley: New York, 2002, 29 pp.
[150] Etzkowitz, H.; Leydesdorff, L., "The Dynamics of Innovation: From National Systems and 'Mode 2' to a Triple Helix of University–Industry–Government Relations," *Res. Policy,* **2000**, *29*, 109–123.

[151] Gordon, R.J., *The Rise and Fall of American Growth*, Princeton University Press: Princeton, NJ, 2016.
[152] Etzkowitz, H.; Leydesdorff, L., "The Triple Helix-University-Industry-Government Relations: A Laboratory for Knowledge Based Economic Development," *EASST Rev.*, **1995**, *14 (1)*, 14–19.
[153] Etzkowitz, H., "Innovation in innovation: The triple helix of university-industry-government relations," *Soc. Sci. Information*, **2003**, *42(3)*, 293–337.
[154] Leydesdorff, L., "The Knowledge-Based Economy and the Triple Helix Model," *Ann. Rev. Information Sci. and Technol.*, **2010**, *44*, 367–417.
[155] Stevens, G.A.; Burley, J., "3,000 Raw Ideas Equals 1 Commercial Success!" *Res. Technol. Mgmt.*, **1997**, *40(3)*, 16–27.
[156] Conference Board, "Trading in the Global Ideas Market. 5th Annual Innovation Report," Report 510-03, Conference Board of Canada: Ottawa, 2002.
[157] Attar, H.; Shahabi, S.M.R., "Beyond 'Funnel' and 'Fireworks': 'Water Ribbed Balloon' as a New Metaphorical Approach to Innovation-in-Practice," *Iranian J. Mgmt. Stud.*, **2014**, *7(1)*, 67–93.
[158] Royce, W., "Managing the Development of Large Software Systems" *Proc. IEEE WESCON*, **1970**, *26(Aug.)*, 1–9.
[159] Bell, T.E.; Thayer, T.A., "Software Requirements: Are They Really a Problem?" Proc. 2nd Internat. Conf. Software Engineering, IEEE Computer Society Press, 1976, pp. 61–68.
[160] Moore, G.A., *Crossing the Chasm*, Harper Collins: New York, 1991.
[161] Johnston, C.; Grant, M., "Innovation Management for Established Businesses. Management Matters," Report 6368, Conference Board of Canada: Ottawa, August, 2014.
[162] Troy, K.L., "Making Innovation Work. From Strategy to Practice," Research Report R-1348-04-RR, Conference Board of Canada: Ottawa, 2004.

Kolibri 360 GmbH
Ferdinand-Jühlke-Straße 7
99095 Erfurt, Deutschland
produktsicherheit@kolibri360.de